McGraw Hill

Physics
Review and Workbook

McGraw Hill

Physics
Review and Workbook

Connie J. Wells

New York Chicago San Francisco Athens London Madrid
Mexico City Milan New Delhi Singapore Sydney Toronto

1 2 3 4 5 6 7 8 9   LHS   27 26 25 24 23 22

ISBN   978-1-264-26408-7
MHID   1-264-26408-9

e-ISBN   978-1-264-26407-0
e-MHID   1-264-26407-0

McGraw Hill products are available at special quantity discounts to
use as premiums and sales promotions or for use in corporate training
programs. To contact a representative, please visit the Contact Us pages
at www.mhprofessional.com.

McGraw Hill is committed to making our products accessible to all
learners. To learn more about the available support and accommodations
we offer, please contact us at accessibility@mheducation.com. We also
participate in the Access Text Network (www.accesstext.org), and ATN
members may submit requests through ATN.

Contents

·1·
Review of Mathematical and Graphing Skills for Physics

·2·
Kinematics

·3·
Forces and Newton's Laws

·15·

Atomic and Nuclear Physics   223

·A·

Fundamental Constants and Useful Information

·B·

Metric Units and Conversions

·C·

Answers to Chapter Exercises

Introduction

This text provides a review of major topics in introductory physics with workbook pages that provide context for each topic and applications of important physics skills. Keep in mind that this is an overview of topics, so a comprehensive full-volume textbook should be used for reference when greater depth of explanation is required.

Topics are organized by chapter, according to the order used in most major textbooks. In this way, there is some scaffolding throughout, with concepts and skills building as the student works through in order. Each topic chapter includes example problems with detailed solutions followed by a workbook section that includes common problem types related to the topic and tasks related to the topic that apply concepts and skills used in physics courses and physics assessments.

Embedded throughout are text boxes with reminders and helpful hints to emphasize common errors or techniques that will be helpful in problem-solving and on tests or quizzes. These deserve special attention.

Chapter 1 is an important review and self-evaluation for students to find areas related to mathematical background that will be essential for problem-solving in physics. If weak areas are determined, additional work may be necessary prior to starting work in subsequent chapters. This chapter also may be used by teachers to evaluate problem areas for students to assist them toward success.

Chapter 3 on forces is noticeably longer than other chapters, with many more worked examples. This chapter addresses two concerns about forces: (1) Concepts related to forces are essential to understanding in most, if not all, topics in physics, so a solid understanding of forces, Newton's laws, and the construction of force diagrams helps in the understanding of other topics of study. (2) Students often have misconceptions about forces that hinder not only their understanding of forces but their understanding of any topic where forces are applied.

Chapter 15 on atomic and nuclear physics contains a quick overview of topics that are often not covered in an introductory course but may answer some questions students encounter during their study of other chapters. Much of this material is part of a chemistry course that often precedes a physics course in high school.

The topics and skills in this review and workbook are also correlated with the Next Generation High School Science Standards (NGSS). These correlations are notated on the first page of each relevant chapter. For more information, go to https://www.ck12.org/ngss/high-school-physical-sciences.

The solutions to chapter exercises are provided in Appendix C with brief explanations for the correct answers.

McGraw Hill

Physics
Review and Workbook

Review of Mathematical and Graphing Skills for Physics

·1·

■ Geometry and Trigonometry Formulas and Relationships
■ Checking Algebra Skills
■ Graphing

Geometry and Trigonometry Formulas and Relationships

Basic Geometry Facts

When two parallel lines are intersected by a transversal, the following angles are equal in measure:

- Alternate interior angles (e.g., $c = b$ and $g = f$ and $a = d$ and $e = h$)

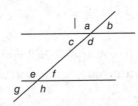

- Corresponding angles (e.g., $f = b$ and $g = c$ and $e = a$ and $h = d$)
- Vertical angles (e.g., $e = h$ and $a = d$ and $g = f$ and $c = b$)

The sum of the measures of the angles of a triangle is 180°.

Area of a rectangle: $A = bh$:

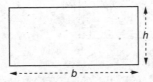

Area of a triangle: $A = \frac{1}{2}bh$:

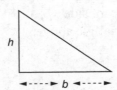

NGSS HS-PS2-1

Area of a circle: $A = \pi r^2$
Circumference of a circle: $C = 2\pi r$

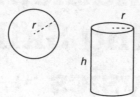

Volume of a cylinder: $V = \pi r^2$
Surface area of a cylinder: $A = 2\pi r^2 + 2\pi rh$
Volume of a sphere: $V = \frac{4}{3} \pi r^3$

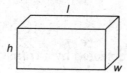

Surface area of a sphere: $A = 4\pi r^2$

Volume of parallelepiped: $V = lwh$

Right Triangle Trigonometry

Consider right triangle ABC, with acute angles A and B and right angle C. The respective sides have length a, b, and c as labeled on the diagram. Sides a and b are called "legs" and side c is called the hypotenuse.

$$a^2 + b^2 = c^2 \quad \text{(Pythagorean Theorem)}$$

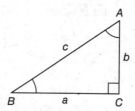

Memorization device for right triangles: SOH-CAH-TOA
sin angle = opposite side /hypotenuse
cos angle = adjacent side/hypotenuse
tan angle = opposite side/adjacent side
For example, from the right triangle above:
sin angle $B = b/c$
cos angle $B = a/c$
tan angle $B = b/a$

1. Sine angle *A* equals:

   A. 29/21
   B. 21/29
   C. 1.05
   D. 8

2. Using triangle *ABC* shown above, find the base length *AB*.

   A. 8
   B. 20
   C. 16
   D. 2.8

3. A flag pole that is 8 m tall casts a shadow on the ground that extends 6 m along the ground from the base of the flag pole. What is the distance from the top of the flag pole to the end of the shadow?

   A. 10 m
   B. 6 m
   C. 4 m
   D. 2 m

4. Referring to the flag pole in the preceding question, what angle does the imaginary line from the tip of the shadow to the top of the flag pole make with the ground?

   A. 30°
   B. 37°
   C. 45°
   D. 53°

5. A streetlight that is 6 m tall shines on a 2-m-tall post standing 3 m from the base of the streetlight. How long is the shadow of the post *x*? (This picture, not drawn to scale, shows the situation.)

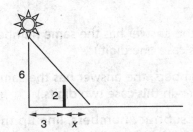

   A. 6 m
   B. 3 m
   C. 2 m
   D. 1.5 m

Checking Algebra Skills

Quadratic Formula: When a quadratic equation is written in standard form $ax^2 + bx + c = 0$,

$$x = -b \pm \sqrt{\frac{b^2 - 4ac}{2a}}$$

Multiplying Binomials (FOIL): Multiply the First terms, the Outer terms, the Inner terms, then the Last terms. To factor, do the opposite.

For example: $(2x + 3)(3x + 4) = 6x^2 + 8x + 9x + 12$ or $6x^2 + 17x + 12$

The following exercises provide practice in basic operations in algebra. *It is important when solving physics problems to follow these steps:*

- Think about the situation given and make a drawing if it helps to picture what is going on.

- Write down what is given and what you are trying to find.

- Find an equation, if possible, that uses the variables you are given.

- Substitute the values given. Pay attention to units and consider whether any units need to be converted in order to yield an answer in proper units.

- Solve the problem, using a calculator if necessary. Facility with a calculator is an important skill in physics.

- Provide the answer with correct units (unless the answer is a variable, which won't have units).

- Think about whether the answer is a reasonable solution for the problem. For example, if you are solving for the speed of a car, an answer in milliseconds might not be reasonable—nor would an answer such as 10,000 m/s.

In Physics, the metric system is most commonly used, so some questions here check your ability to convert among metric units and to judge whether metric answers are reasonable. Appendix A may be helpful.

In some Physics courses and in the laboratory, it is important to provide measurements and answers with the correct number of significant digits. Here are some rules for mathematical operations with numbers and rounding to a reasonable number of significant digits:

EXAMPLES

Ex 1. $2 \times 2 = 4$, not 4.00 (*Rule:* When you multiply numbers, the answer has the same number of significant digits as the factor with the least number—in this case one digit.)

Ex 2. $6.4/3.21 = 2.0$, not 1.993769 (*Rule:* When you divide numbers, the answer has the same number of significant digits as the factor with the least number—in this case two digits.)

Ex 3. $3.275 - 2.4 = 0.9$, not 0.875 (*Rule:* When you add or subtract numbers, line up the decimal points, and round the answer to the least number of decimal places.)

Ex 4. The average of 3.28, 6.059, and 1.0073. The answer is 3.45. (*Rule:* Follow the rule above, but consider the number 3, which you divided, as being an exact number, so it has an infinite number of significant digits.)

1. Solve for x in the following equation: $6(x + 1) = 14x - 6$

   A. 0
   B. 0.625
   C. 0.875
   D. 1.5

2. Solve for x in this equation: $x^2 = 6x$

   A. $x = 0$ or 4
   B. $x = 0$ or -0.25
   C. $x = 0$ or 6
   D. $x = 0$

3. Solve for x in this equation: $2x^2 - x = 0$

   A. $x = -1$ or -3
   B. $x = -1.73$ or 1.73
   C. $x = -0.95$ or 1.45
   D. $x = 0$ or 0.5

4. Solve for t in this equation: $2t^2 - 6t = 0$

   A. $t = 0$ or 3
   B. $t = 0$ or -1
   C. $t = 3$
   D. $t = 0$

5. Solve for r in this equation: $F = mv^2/r$

   A. F/m
   B. mv^2/F
   C. Fmv^2
   D. F/mv^2
   E. Fvm

6. How many micrometers are in one centimeter?

   A. 10^{-6}
   B. 10^{-4}
   C. 10^2
   D. 10^4

7. Which of the following is the closest approximation of 60 feet per second in meters per second?

   A. 18 m/s
   B. 30 m/s
   C. 60 m/s
   D. 100 m/s

The following variables are listed with their units:

Variable	Unit
x	meters
F	kilogram $\cdot$ meters/second^2
v	meters/second
t	seconds
m	kilogram
a	meters/second^2

8. In which of the following equations are the units <u>not</u> balanced?

   A. $x = vt$
   B. $Ft = mv$
   C. $x = at^3$
   D. $F = ma$

9. While driving in Europe, you see a sign indicating that the speed limit is 120. What are the most likely units for this speed?

   A. kilometers/hour
   B. meters/second
   C. miles/second
   D. miles/hour

10. If light travels at 186,000 miles per second, how far, in meters, is a light year (the distance light travels in one year)?

   A. 3×10^8 m
   B. 5.87×10^{12} m
   C. 9.46×10^{15} m
   D. 6.79×10^7 m

11. A student group obtains the following measurements for the length of the same object: 2.43 m, 2.091 m, 2.5 m, and 2.710 m. What average should they report?

   A. 2.425 m
   B. 3 m
   C. 2.4 m
   D. 2 m

> A strong recommendation: When solving problems, use unit analysis to check your work and detect mathematical or unit conversion errors.

12. In a pendulum experiment, students obtain a value of 9.76 m/s^2. Calculate the percent error on the students' experimental value for g, if the accepted value is 9.81 m/s^2.

   A. 5.0%
   B. 0.005%
   C. 0.5%
   D. 0.05%

13. To the correct number of significant digits, what is the sum of 12.035 and 1.75 and 6.301?

    A. 20.09
    B. 20.086
    C. 20.1
    D. 20

14. Using our rules for determining significant digits on answers to mathematical operations, what would you report for the square root of 4.0?

    A. 2.000
    B. 2.00
    C. 2.0
    D. 2

15. Solve for v in the equation $K = \frac{1}{2} mv^2$.

    A. $\dfrac{2K}{m}$

    B. $\sqrt{\dfrac{2K}{m}}$

    C. $\sqrt{2Km}$

    D. $2Km$

16. Convert 35 miles per hour to meters per second.

    A. 40.2 m/s
    B. 12.1 m/s
    C. 78.3 m/s
    D. 15.6 m/s

17. Which statement uses prefixes correctly?

    A. 100 cm = 10 km
    B. 2.5 kg = 2500 cg
    C. 1,000 ml = 1 dl
    D. 150 cm = 1.50 m

18. 100 meters is <u>not</u> equivalent to:

    A. 1.00×10^4 cm
    B. 1.00×10^{-1} km
    C. 1.00×10^2 m
    D. 1.00×10^6 mm

Graphing

Graphical analysis is one of the most important tools we have in Physics to make sense of our observations. By examining graphically the data from experiments, one can derive equations to describe the relationships between the variables in the experiments.

General form of a line: $y = mx + b$, where y is the value on the y-axis or vertical axis, x is the value on the x-axis or horizontal axis, m is the slope of the line, and b is the y-intercept of the line.

Slope of a line of form $y = mx + b$: $m = \dfrac{\text{rise}}{\text{run}} = \dfrac{\Delta y}{\Delta x} = \dfrac{y_2 - y_1}{x_2 - x_1}$

Independent variables: Data values manipulated by the experimenter, usually recorded in the first column of a data table and on the horizontal or x-axis.

Dependent variables: Data values that change as a result of changes in the independent variable, usually recorded in the second column of a data table and on the vertical or y-axis.

Part I. Linear Graphs

This plot of "y versus x" is a direct, linear relationship of the form $\mathbf{y = mx + b}$, where m is the **slope** and b is the **y-intercept** (which is 0 in this case).

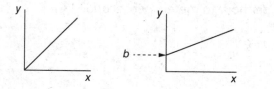

This plot of "y versus x" is a direct, linear relationship of the form $\mathbf{y = mx + b}$, where $\mathbf{m}$ is the slope and $\mathbf{b}$ is the y-intercept.

In an experiment where the **independent variable** is defined by x and the **dependent variable** is defined by y, the graph should be set up with the independent variable on the x-axis and the dependent variable on the y-axis so that we have graphed "y versus x," or "y as a function of x."

> As soon as you see a linear graph on an assignment or on a test, ask yourself: What do the slope, y-intercept, and area under the graph line mean? ("Under the graph line" means between the graph line and the positive x-axis.)

Experiment 1: In the laboratory, students gather the following data from an experiment in which the distance d a small car moves along a surface is measured as a function of the time the car has traveled. The students construct a data table with the independent variable, t, listed on the left and the dependent variable, d, listed on the right (which is the usual convention):

Time, t (s)	Distance, d (m)
0	0
1.0	1.3
2.0	2.4
3.0	4.0

The students produce Graph 1, keeping the following elements of a graph in mind:

1. Draw the axis lines using a straight-edge.

2. Label each axis with quantity, symbol, and units.

3. Give the graph a title of form "y versus x."

4. Number each axis at regular intervals, starting at zero, and using a range of numbers defined by the range of values on the data table so that the plot fills the space.

5. Draw a best-fit line, using a straight-edge, that fits most values—but does not necessarily go through all the data points.

6. Draw the best-fit line through the point (0, 0) only if those values fit the real situation.

7. When evaluating the slope using "rise over run" or "change in y-value divided by change in x-value," use only points that are on the best-fit line. Do not go back to the data table to use points in the data table. Use actual data points only if they are on the best-fit line.

8. Give the slope units that are defined by the units on the graph, i.e., units on y-axis divided by units on x-axis.

Graph 1

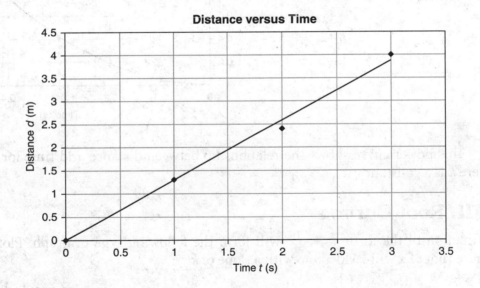

Analysis of Graph 1

The graphical plot is linear, so distance is directly and linearly proportional to time. The y-intercept of the line is at zero, so distance is zero when time is zero. Substituting into the linear form $y = mx + b$, $d = mt + 0$, or $d = mt$, where m is the slope.

Finding the slope, m, for the line (which is the same at every point, since it is a line), $m = \Delta y / \Delta x$. Any points *on the line* may be chosen, but here it's easy to use the points (0.75, 1) and (0, 0). Then $m = 1/0.75 = 1.3$ m/s. Substituting again, using this slope, $d = (1.3$ m/s$)t$. This equation can be checked by applying it to other points on the line.

Part II. Power Curves

We will restrict our study of power curves to those that have an equation of the form $y = kx^2$, where k is a constant. The graph of this relationship generally looks like the diagram below. A plot of y versus x^2 will produce a line with a slope of k.

Experiment 2. A group of students tie a string to a cart with a mass hanging from the edge of the table. As the falling mass pulls the cart, accelerating it across the table, the students acquire the following data for the cart's distance traveled as a function of time and then produce Graph 2.

Graph 2

Time, t (s)	Distance, d (m)
0	0
0.1	0.045
0.2	0.187
0.3	0.423
0.4	0.764
0.5	1.190

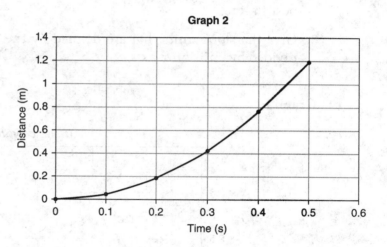

This power curve shows the relationship between distance and time for the cart: $D = kt^2$, where k is a constant.

Part III. Root Curves

A function of the form $y = k\sqrt{x}$ will form the following type of graph. Plotting y versus the square root of x will yield a line with a slope of k.

Determining the value of k to find the equation for a graph of this type involves a process similar to that used in Part II. You can see that a graph of "y versus square root of x" would be a direct proportion and would produce a line. Find the square roots of all the x-values and plot again, with the y-values on the y-axis and the square root of x-values on the x-axis. It will make a line, so find the slope of that line to determine k.

Experiment 3.   A group of physics students are asked to determine a value for *g* using a swinging bowling ball pendulum. They take the following measurements and make Graph 3 that you see below.

Graph 3

Length of pendulum, *L* (m)	Period of swing, *T* (s)
0	0
0.2	0.90
0.4	1.27
0.7	1.68
0.9	1.90
1.2	2.20

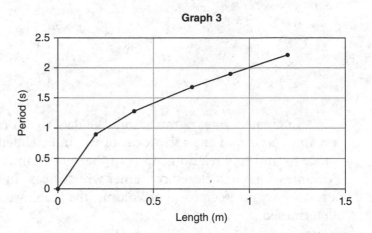

The students then make a new data table that shows what they will plot to create a line. The original graph above is a "root curve" showing the relationship between Period and the square root of Length. Therefore, a plot of Period on the *y*-axis and square root of Length on the *x*-axis should produce a line, as shown in Graph 4.

Graph 4

Square root of Length	Period
0	0
0.45	0.9
0.63	1.27
0.84	1.68
1.38	1.90
1.48	2.20

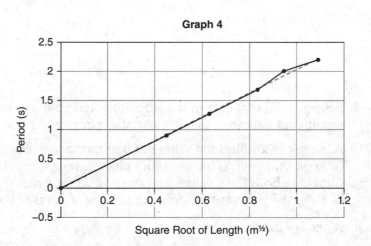

The best-fit line is shown as a pale dashed line on the graph. Find the slope of the best-fit line, which is:

$$\text{Slope} = \frac{1-0}{0.5-0} = 2\frac{\text{s}}{\sqrt{\text{m}}}$$

Since the equation for a pendulum is $T = 2\pi\sqrt{\dfrac{L}{g}}$ and the students have plotted T versus $\sqrt{L}$, then the slope must be equal to $\dfrac{2\pi}{\sqrt{g}}$. Setting this equal to the slope, the students obtain 9.87 m/s^2 as the value for *g*.

Part IV. Inverse Curve

Inverse curves describe an inverse relationship between two variables: $y = k(1/x)$. In other words, this type of relationship says that as the value of y increases, the value of x decreases. The graph is of the following form:

To create a linear graph from the data that forms this graph, plot the dependent variable on the y-axis and the reciprocal of the independent variable $(1/x)$ on the x-axis. A common place to find this relationship is in thermodynamics, where we plot Pressure as a Function of Volume. A gas in a closed container will decrease in volume as pressure increases. If you plotted Pressure versus Reciprocal of Volume, the graph would be a line, from which the slope could be determined.

EXERCISE 1.3

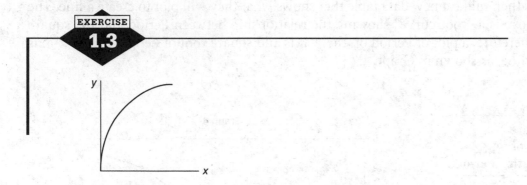

1. Suppose you plot data in the laboratory and obtain a curve as shown here. How would you replot your data to obtain a linear graph from which you could determine a slope?

   A. Keep the x-values the same but plot the square root of the y-values on the y-axis.
   B. Keep the y-values the same but plot the square root of the x-values on the x-axis.
   C. Square both the x-values and the y-values and plot them.
   D. Keep the y-values the same but plot the squares of the x-values on the x-axis.

Questions 2–4 use this graph.

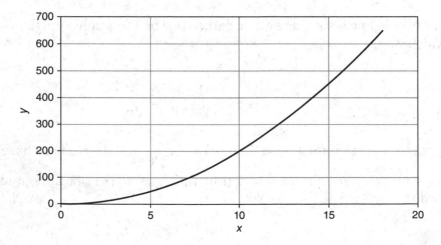

2. Suppose you plot data in the laboratory and obtain a curve as shown here. How would you replot your data to obtain a linear graph from which you could determine a slope?

   A. Keep the x-values the same but plot the square root of the y-values on the y-axis.
   B. Keep the y-values the same but plot the square root of the x-values on the x-axis.
   C. Square both the x-values and the y-values and plot them.
   D. Keep the y-values the same but plot the squares of the x-values on the x-axis.

3. Identify the equation for the graph given previously:

   A. $y = 2\,x^2$
   B. $y = 20\,x$
   C. $x = 0.2y$
   D. $y = \frac{1}{2}\,x$

4. If the graph given previously has the title of "Distance versus Time," what would be the distance when time is equal to 10 s?

   A. 10 m
   B. 100 m
   C. 200 m
   D. 300 m

5. Of the four graphs below, which one best represents y versus x^2?

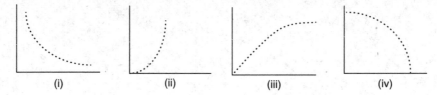

| (i) | (ii) | (iii) | (iv) |

   A. (i)
   B. (ii)
   C. (iii)
   D. (iv)

6. On a linear graph that has a quantity measured in newtons on the y-axis and a quantity measured in meters on the axis, how would you use units to identify the quantities identified by the slope and by the area under the graph?

   A. The slope would have units of N/m, and the area would have units of N-m.
   B. The slope would have units of N-m, and the area would have units of N/m.
   C. The slope would have units of N, and the area would have units of N-m.
   D. The slope would have units of m/N, and the area would have units of N-m.

Kinematics

- **Vectors and Scalars**
- **Position, Displacement, and Constant Velocity**
- **Constant Acceleration**
- **Graphical Analysis of Motion**

Kinematics: This is the study of how objects move. In this chapter, you will review terms that describe motion, units used to measure them, and the difference between vector and scalar quantities. You will also review motion graphing as a method of analyzing motion.

Kinematic Equations: This is a set of equations that are used most commonly when working problems in kinematics:

$$\Delta X = V_x t + \frac{1}{2} a t^2$$
$$V_f = V_o + at$$
$$V_f^2 = V_o^2 + 2a\Delta X$$

Linear Motion: Also called *translational motion*; when the center of mass of an object under consideration moves from one position to another without rotation.

Vectors and Scalars

Scalar: Describes a quantity that has magnitude but not direction. Some scalar quantities include distance, speed, time, energy, and temperature.

Vector: Describes a quantity that has both a size (magnitude) and direction. Some vector quantities include position, displacement, velocity, acceleration, and momentum. In this book, symbols of vectors will either be typed in bold or have a vector symbol written above.

When adding vectors, add them "tip to tail" as shown here, where Vector **A** is added to Vector **B**. The **resultant** is the vector drawn from the beginning of the first vector to the end of the last vector.

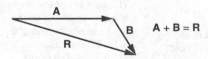

NGSS HS-PS2-1

15

When subtracting vectors, add the negative of the subtracted vector, as shown below. The negative of a vector is the same size but opposite in direction.

Vectors in two dimensions: These are vectors that are just in opposite directions. For example, a vector that is a velocity 10 m/s east could be added to a velocity vector that is 6 m/s west. We would make one positive and one negative. So the first one is +10 m/s and the second one is −6 m/s. If we add them, we get +4 m/s, which is 4 m/s east.

Perpendicular Vectors: When two vectors are added at right angles to each other, the **Pythagorean Theorem** can be used to find the resultant. The vectors will combine as the legs of a right triangle, and the resultant will be the hypotenuse.

EXAMPLE—Vector Addition

Finding a displacement for vectors in two dimensions using only positive and negative for direction:

A cart on a track starts at $x = 0$ and moves a distance of 2.5 m in the positive direction, then bounces off a spring and moves 1.5 m in the negative direction before stopping. What is the displacement of the cart during this experiment?

Solution:

Add the two vectors with their signs: net $\Delta x = +2.5$ m $+ (−1.5$ m$) = 1.0$ m

EXAMPLE (Advanced)—Vectors at Right Angles

Finding a resultant for vectors at right angles.

Add a vector of 20 m to the north to a vector of 10 m to the east.

Solution:

Use the Pythagorean theorem to find the hypotenuse, which is the magnitude of the resultant:

$$R^2 = (20\,\text{m})^2 + (10\,\text{m})^2$$
$$R = \sqrt{(400+100)} = 22.4\,\text{m}$$

We can now use some trigonometry to find the direction of the resultant.

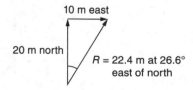

The tangent of the angle marked is equal to 10/20, so the angle is the inverse tangent of 0.5, which is 26.6°. The resultant is directed at 26.6° east of north.

1. Which answer choice is the resultant of vector **A** minus vector **B**?

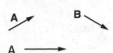

   A. ———►
   B. ◄———
   C. ↑
   D. ↓

2. Which answer choice is the resultant of vector **A** plus vector **B**?

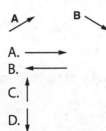

   A. ———►
   B. ◄———
   C. ↑
   D. ↓

3. An airplane is flying north with airspeed 200 km/h when it meets a crosswind of 70 km/h toward the east. The expression to determine the resultant speed of the airplane is:

   A. $(200 + 70)$ km/h
   B. $(200 - 70)$ km/h
   C. $\sqrt{(200^2 + 70^2)}$ km/h
   D. $\sqrt{(200^2 - 70^2)}$ km/h

4. An airplane starts on a course due north at an airspeed of 100 km/h and encounters a crosswind from the west at 10 km/h. Which is the most likely resultant flight path of the airplane?

   A. 110 km/h on a path due northwest
   B. 110 km/h on a path due northeast
   C. 101 km/h on a path east of north
   D. 101 km/h on a path west of north

5. A student walks 100 m west and then 32 m east. If east is considered the positive direction, what is the displacement of the student at the end of the walk?

   A. 64 m
   B. −64 m
   C. 132 m
   D. −132 m

Position, Displacement, and Constant Velocity

Position: The location relative to a starting point or a reference point. For example, an object's position may be at the origin of a grid, which is at (0, 0), or it may be located at a position 2.0 m away from the origin at an angle of 30° north of east. The symbol may be x or y and is usually measured in meters.

Distance: A scalar quantity that answers the question "How far?" Distance traveled has no indication of direction.

Displacement: A vector quantity that describes a change in position, including a distance and a direction. If a positive displacement is in a given direction, then negative displacement is in the opposite direction.

For example, an object may be displaced a distance of 2.0 m at an angle of 30° north of east from its original position to its final position: $\Delta x = x_{final} - x_{original}$:

$$\Delta x = v_o t + \frac{1}{2} a t^2$$

Speed: Answers the question "How fast?" Speed is a scalar quantity, v, usually measured in meters per second (m/s):

$$v = d/t$$

Velocity: Answers the questions "How fast and in what direction?"
Velocity is a vector quantity that has both speed and direction, so it is measured in meters per second (m/s) in a specified direction. Change in velocity, $\Delta \mathbf{v}$, is equal to $\mathbf{v}_{final} - \mathbf{v}_{original}$. When velocity is just being considered in two dimensions, it is convenient to call velocity in one direction positive and velocity in the opposite direction negative.

Average Velocity: The average of velocities over a time period or displacement divided by time.

$$\mathbf{V}_{average} = \frac{\mathbf{V}_{final} + \mathbf{V}_{original}}{2}$$

$$\mathbf{V}_{avg} = \frac{displacement}{time} = \frac{\Delta x}{t}$$

Instantaneous Velocity: The average velocity of an object during an infinitesimally small time interval; that is, the change in position of a moving object during a time interval that approaches zero. If the object is moving at constant velocity during the entire time under consideration, then the average velocity and instantaneous velocity are the same.

Center of Mass: Point where the mass of an object is concentrated, or "balance point" of an object or system in a gravitational field. (See the first section of Chapter 6 for calculation.)

Velocity, v: A vector measurement of the rate in change of position or rate of displacement.

$$\mathbf{v} = \frac{\Delta x}{t} = \frac{x_f - x_o}{t}$$

$$\mathbf{v}_f = \mathbf{v}_o + at$$

(Acceleration in one dimension and acceleration in two dimensions in terms of projectile motion are discussed in Chapter 4.)

A change in any quantity is always "final minus initial." Remember to put the proper signs on each vector to indicate direction. For example, a change in velocity from +2**v** to −**v** is equal to −**v** minus 2**v** or −3**v**.

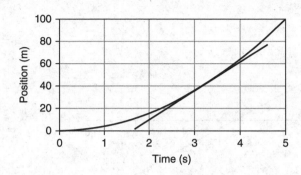

On the graph above, the tangent line at $t = 3$ s is drawn. The slope of that tangent line (about 15 m/s) is the instantaneous velocity at 3 s.

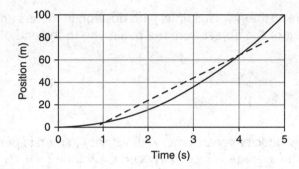

On a graph of Position versus Time, average velocity is the slope of the line connecting two points. For example, to find the average velocity between $t = 1$ s and $t = 4$ s, the solid line is constructed. The slope of the dashed line would be the average velocity between those two points (about 20 m/s).

EXERCISE 2.2

1. The average velocity for a trip in which an object travels a distance of 20 m in 10 s:

   A. might be equal to zero
   B. must be equal to 2 m/s
   C. might be greater than 2 m/s

2. A student walks completely around an oval ¼-mile track in 3 minutes. What is the student's displacement for this trip?

   A. zero
   B. ¼ mi
   C. ¾ mi
   D. 1,600 m

3. A student walks completely around an oval ¼-mile track in 3 minutes. What is the student's average velocity for this trip?

   A. zero
   B. $\frac{1}{12}$ mi/min
   C. ¾ mi/min
   D. ½ mi/min

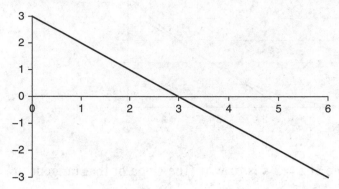

4. If the above plot is position versus time, with position in meters and time in seconds, what is the displacement of the object from the beginning to the end of 6 s?

   A. 6 m
   B. −6 m
   C. zero
   D. −3 m

5. If the above plot is velocity versus time, with velocity in meters per second and time in seconds, what is the average velocity between $t = 2$ s and $t = 4$ s?

   A. zero
   B. −1 m/s
   C. +1 m/s
   D. −2 m/s

Constant Acceleration

Acceleration: Change in velocity over a given time period. It is a vector quantity measured in meters per second per second (i.e., velocity units divided by a unit of time). Negative acceleration generally means a decrease in velocity in the positive direction or an increase in velocity in the negative direction. Positive acceleration generally means an increase in velocity in the positive direction or a decrease in velocity in the negative direction.

$$\mathbf{a} = \frac{\mathbf{V}_{final} - \mathbf{V}_{original}}{\Delta t} = \frac{\Delta \mathbf{V}}{\Delta t}$$

Instantaneous Acceleration: Acceleration over an infinitesimally small time period (i.e., $\bar{a}_{ave}$ as $t \rightarrow 0$). On a graph of velocity versus time, it is the slope of the tangent at a specified point.

Average Acceleration: On a graph of acceleration versus time, average acceleration is the slope of the line connecting two points.

Gravitational acceleration: For a **freely falling object** near Earth's surface, the acceleration is g, or 9.8 m/s^2, toward the center of Earth. Mass of the object does not affect the gravitational acceleration of a falling object (if air resistance is negligible).

> A common mistake students make in problem-solving is automatically making g negative. The negative sign is only an indicator of direction and is a conscious decision made by the problem solver. For example, if you're working a problem for an object falling downward, you might make the displacement negative and the velocity negative and then the acceleration, g, negative.

Falling objects: Objects moving straight upward or falling straight downward are in **free fall** and have a constant downward acceleration of **9.8 m/s^2**, or g, near the Earth's surface. Objects in projectile motion are also in free fall and have only a constant acceleration downward of g. We assume in both cases that air friction on the objects in free fall is negligible.

Galileo's Law of Odd Numbers: Galileo theorized and showed by experiment that if acceleration is constant, an object will move distances in each subsequent time interval that follow the odd numbers. For example, if the object moves a distance x in the first second, it will move a distance $3x$ in the second and a distance of $5x$ in the third second, and so on.

Velocity versus Time

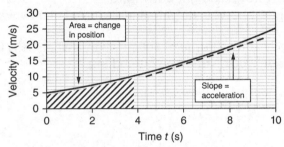

☆ Use the units to help you confirm whether to take slope or area. In the case at the left, slope will have units of m/s over seconds—so that is acceleration.

EXERCISE 2.3

1. An object starts from rest at $x = 0$ m and accelerates at a constant rate, moving from a position of $x = 0$ m to $x = 2$ m in 4 s. How far will the object move in the next 4 s?

   A. 3 m
   B. 4 m
   C. 5 m
   D. 6 m

2. The slope of a point on a Velocity versus Time graph is:

   A. change in position
   B. average acceleration
   C. instantaneous acceleration
   D. change in velocity

3. A plot of Velocity versus Time for a moving object is a straight line. Which of the following can be true?

   A. The velocity is constant.
   B. The acceleration is constant.
   C. The acceleration and velocity are both zero.
   D. All of the above could be true.

4. Which of the following best describes the acceleration and velocity vectors for an air rocket fired straight upward from the ground, from the moment just after it leaves the ground until it hits the ground again? Assume the upward direction to be positive.

   A. The acceleration is always negative, and the velocity is positive on the way up and negative on the way down.
   B. The acceleration and velocity vectors are both positive on the way up and both negative on the way down.
   C. The acceleration and velocity vectors are both positive on the way up and on the way down.
   D. The acceleration is always negative, and the velocity is always positive.

5. A stone thrown straight upward without air friction has an acceleration that is:

   A. smaller than that of a stone thrown straight downward
   B. the same as that of a stone thrown straight downward
   C. greater than that of a stone thrown straight downward
   D. zero until it reaches the highest point in its path and starts to fall downward

6. An airplane is flying horizontally at an altitude of 500 m when a wheel falls from it. If there were no air resistance, the wheel would strike the ground in:

   A. 10 s
   B. 20 s
   C. 50 s
   D. 80 s

7. A baseball is hit so that it rises almost vertically into the air. From the time it is hit until the ball reaches the level of the baseball bat again, an observer counts a total time of 4 s. What is the best estimate of the velocity of the ball as it left the baseball bat after it was hit?

   A. 10 m/s
   B. 20 m/s
   C. 30 m/s
   D. 40 m/s

Graphical Analysis of Motion

Graphical analysis is an effective way of describing the motion of an object. Here is a summary of what these graphs mean, with exercises to follow that apply these rules.

I. Position versus Time Graphs:

- Slope is the velocity during that time interval if the graph is a line. If the graph is a curve, the slope of the tangent line at a point on the curve is the velocity. (See Diagram B on the next page.)

- Straight lines represent constant velocity, with positive slopes representing velocity in the positive direction and negative slopes representing velocity in the opposite, or negative, direction. A line with no slope represents zero velocity, or no change in position.

- Curved lines indicate a changing velocity, or acceleration.

Diagram A

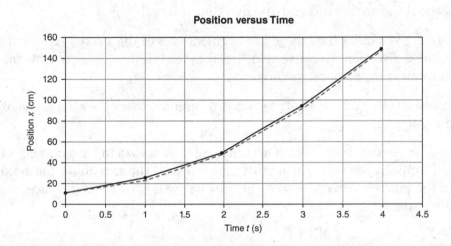

Example for a cart accelerating along a track:

1. The cart starts at a position of 10.5 cm when $t = 0$.

2. The position keeps increasing in the positive direction, so movement along the track is considered positive and the cart never turns around.

3. The position is increasing at an increasing rate, so it is **accelerating**.

4. We might conclude that a graph of position versus time that curves upward like this shows an object that is accelerating.

5. The slope at every point on the graph is the **velocity,** since slope equals rise over run, and that would be $\Delta x / \Delta t$ or **v**, in cm per second. Diagram B (next page) has a tangent drawn for several data points on the position versus time graph. The slope of each tangent is the velocity at each of those points. Those are the velocities that are plotted on the Velocity versus Time graph (Diagram C).

6. The **displacement** of the cart for the 4 s shown here is the final position at 4 s, which is 150 cm, minus the initial position at 0 s, which is 10 cm. Thus, the displacement is 140 cm.

Diagram B

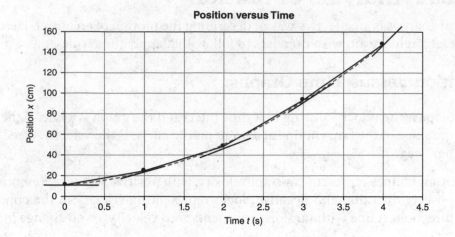

Position versus Time

II. Velocity versus Time Graphs:

- Slope is the acceleration during that time interval.

- Positive velocities (i.e., those line segments above the x-axis) represent movement in the positive direction. Negative velocities (i.e., those line segments below the x-axis) represent movement in the negative direction.

- When the line on a velocity versus time graph crosses the x-axis, the object has changed direction.

- The area under a velocity versus time graph (as shown in Diagram D) is the displacement, or change in position, during that time interval. The area above the x-axis is displacement in the positive direction, and the area below the x-axis is displacement in the negative direction.

Diagram C

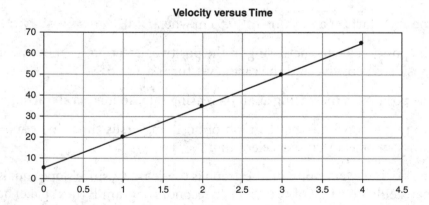

Velocity versus Time

Diagram D

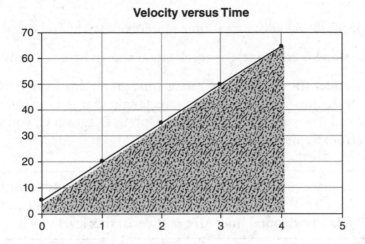

Velocity versus Time

Example of Velocity versus Time graph for the *X* versus *T* graph from Part I:

1. Since the velocity is increasing, the object must be accelerating.

2. The slope is constant and positive, so the object has a positive acceleration that is equal to "rise over run," which is 60 m/s over 4 s, or 15 m/s/s.

3. The area under the graph line is the *change in position* (as shown by the shaded area above). The area under the Velocity versus Time graph is the displacement of the cart (or distance it traveled down the track). The total area between the graph line and the *x*-axis in this case includes a large triangle and a smaller rectangle. Area = ½ base × height + base × height = ½ × 4 × 60 cm + 5 × 4 = 140 cm. (See Diagram D above.)

III. Acceleration versus Time Graphs:

- The area under a graph line segment on an **acceleration versus time** graph is the change in velocity.

- The slope of an acceleration versus time graph is often described as jerk, which is change in acceleration. At this level we are dealing only with constant acceleration situations, so all of our acceleration versus time graphs will be horizontal lines.

Diagram E

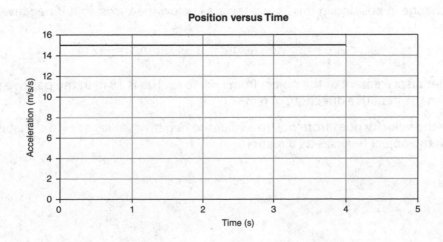

Position versus Time

Example of acceleration versus time graph for the situation described in the previous graphs in this section:

- The slope of the velocity versus time graph is constant and equal to 15 m/s².

- Since the slope is constant, the acceleration shown here is a constant value.

- The area under the acceleration versus time graph line from $t = 0$ to $t = 4$ s is the *change in velocity*. In this case, the area is a rectangle that is 15 m/s/s by 4 s, which is 60 m/s. (If you check the velocity versus time graph in Diagram C, you will see that the velocity changes from 5 m/s to 65 m/s to confirm this.)

> When the displacement versus time graph is linear, velocity is constant, since the slope of the line is velocity. When the velocity versus time graph is linear, the acceleration is constant, which is a horizontal line. All examples at this level will be constant acceleration.

EXAMPLE—Velocity versus Time Graph

The graph below illustrates the velocity as a function of time for an object. (a) What is the object's instantaneous acceleration at $t = 5$ s? (b) Determine the change in the object's position for the first 6 s. (c) What is the change in position from $t = 6$ to $t = 10$ s? (d) Is the object moving in the same direction during the entire time of the graph? (e) Describe the object's motion.

Solution:

(a) The instantaneous acceleration is the slope of the line at $t = 5$ s. Since this is a line, the slope is the same for the entire line, so we can choose any set of points. Use rise/run for the entire line from $t = 0$ to $t = 6$ seconds:

$$\text{slope} = \frac{\text{rise}}{\text{run}} = \frac{(0-6)\,\text{m/s}}{(6-0)\,\text{s}} = -1\,\text{m/s}^2$$

(b) Change in position, or displacement, is the area under the line from $t = 0$ to $t = 6$ s. This is a triangle:

$$\text{Area} = \tfrac{1}{2}\text{base} \times \text{height} = \tfrac{1}{2}(6\text{ s})(6\text{ m/s}) = 18\text{ m}$$

(c) The change in position in this case is the area below the *x*-axis, so it is negative.

$$\text{Area} = \tfrac{1}{2}\text{base} \times \text{height} = \tfrac{1}{2}(-4\text{ s})(6\text{ m/s}) = -12\text{ m}$$

The net displacement of the object from $t = 0$ to $t = 10$ s is 18 m in the positive direction and 12 m in the negative direction, or 6 m.

(d) Since the velocity goes from positive values to negative values at $t = 6$ s, that is the point at which the object reverses its direction.

(e) The acceleration is the same negative value throughout, so the object is slowing down in the positive direction for the first 6 s and then speeding up in the negative direction for the final 4 s.

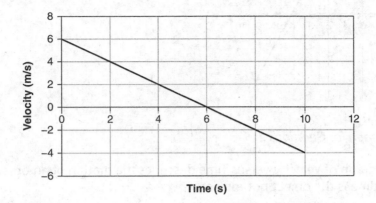

Time (s)

1. Pick the best use of words included in the description of the motion of an object using the following Velocity versus Time graph:

A. The object starts at the origin at $t = 0$ s and moves at a constant speed during the entire time.
B. The object starts at the origin at $t = 0$ s and moves at a decreasing speed.
C. The object starts at the origin at $t = 0$ s and moves away at an increasing speed.
D. The object starts at the origin at $t = 0$ s and first moves away at an increasing speed and then at a decreasing speed.

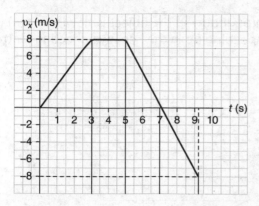

Use the above graph for Questions 2–5.

2. The above graph of Velocity versus Time describes the motion of an object. During which point(s) in time is the object not moving?

   A. From $t = 3$ to $t = 5$ s
   B. At $t = 8$ s
   C. From $t = 7$ to $t = 9$ s
   D. At $t = 0$ and $t = 7$ s

3. When is the object moving in the negative *x*-direction?

   A. From $t = 7$ to $t = 9$ s
   B. From $t = 5$ to $t = 7$ s
   C. From $t = 3$ to $t = 5$ s
   D. The object is always moving in the positive *x*-direction.

4. How far did the object move during the first 3 s?

   A. 2.67 m
   B. 8 m
   C. 12 m
   D. 24 m

5. What is the acceleration of the object during the first 3 s?

   A. 2.67 m/s/s
   B. 8 m/s/s
   C. 12 m/s/s
   D. 24 m/s/s

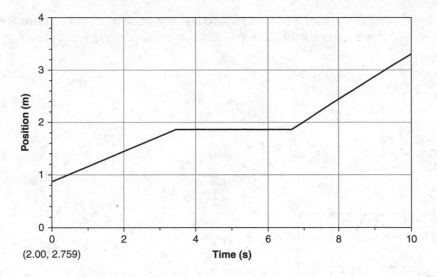

(2.00, 2.759)

Use the above graph to answer Questions 6 and 7.

6. This graph could have been made by a student walking toward or away from a motion sensor in the laboratory. The motion sensor sends out and receives signals to determine the position of the student at each moment in time. What is the displacement of the student from $t = 0$ s to $t = 10$ s?

   A. 0.9 m
   B. 3.3 m
   C. 2.0 m
   D. 2.4 m

7. During which interval of time is the student moving fastest, and why?

   A. From 6.5 to 10 s, because velocity is the slope and the slope is largest in that interval
   B. From 0 to 3.5 s, because the largest displacement occurs in the smallest amount of time
   C. From 3.5 to 6.5 s, because there is no change in position
   D. The student is always walking at the same speed or stopped, since the slopes are always zero or equal to each other.

8. This is a graph of Velocity versus Time for a moving object, where positive displacement is considered forward. Select the answer that lists all points or intervals during which the object is not moving.

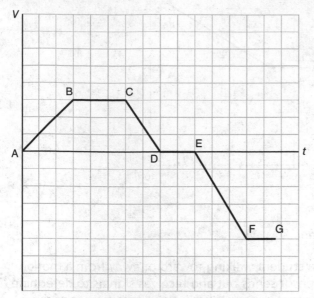

A. Point A, interval BC, interval DE, and interval FG
B. Interval BC and interval FG
C. Point A, interval DE
D. Interval DE only

9. On the above graph of Velocity versus Time for a moving object, positive displacement is considered forward. At what labeled points does the object turn around and move backward?

A. Point B
B. Point C
C. Point E
D. Point F

> When the graph of Velocity versus Time is a line, the slope of that line is acceleration, and the area between the line and the x-axis is the distance traveled during that time period.

10. When you compare two different position versus time graphs on the same graph axes, both having a linear positive slope over the same amount of time, the graph with the steeper slope:

A. Represents the smaller velocity of the two
B. Represents constant position for both, since the slopes are positive
C. Represents a smaller net displacement
D. Represents greater rate of change of position

11. If we had a velocity versus time graph that showed you moving for 4 s at a constant velocity of negative 2 m/s, your net displacement during the 4 s is:

    A. Zero
    B. Negative 2 m
    C. Negative 8 m
    D. Negative 0.5 m

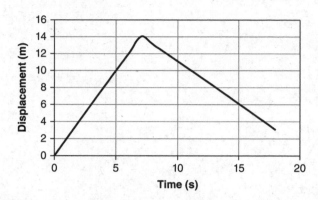

Use the above graph for Questions 12–14.

12. The displacement of the object described in this graph from 0 s to 5 s is:

    A. 5 m
    B. 10 m
    C. 15 m
    D. 20 m

13. The constant slope of this graph from 0 s to 6 s indicates:

    A. no change in position
    B. constant velocity
    C. constant acceleration
    D. constantly changing velocity

14. The average velocity from 10 s to 15 s on the graph is closest to:

    A. 1.5 m/s
    B. −1.0 m/s
    C. zero
    D. −2.0 m/s

15. Which of the following is an example of motion at constant velocity?

    A. An object in free fall after being dropped
    B. An object moving in a circle at constant speed
    C. A car that drives at constant speed up, over, and back down a hill
    D. A box sliding at constant speed down a ramp

When the graph of position versus time is a curve, there is acceleration during that time period. A concave-up (opening upward) curve is an increase in speed, and a concave-down (opening downward) curve is a decrease in speed.

16. The graph below describes the motion of an object as position versus time. What is the best estimate of the velocity of the object at a time $t = 7$ s?

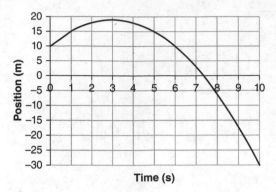

A. 2 m/s
B. −2 m/s
C. 7 m/s
D. −7 m/s

17. Which of the answer choices best describes the motion of an object as depicted on the graph of Position versus Time shown above, assuming the forward direction is positive?

A. The object moves forward for about 7 s, reverses direction at $t = 7$ s, and then moves backward for about 3 s.
B. The object moves forward for about 3 s, reverses direction at $t = 3$ s, and then moves backward for about 7 s.
C. The object speeds up for the first 3 s and then slows down during the time from $t = 3$ s to $t = 7$ s.
D. The object starts its motion by moving forward and then reverses its direction at $t = 3$ s and again at about $t = 7$ s.

Forces and Newton's Laws

·3·

- **Types of Forces and Free-Body Diagrams**
- **Newton's Laws**
- **Equilibrium**

Forces can change the motion of objects and are defined as a push or a pull on an object or system. Isaac Newton carefully defined three laws that describe the nature of forces using what he knew at the time about how forces behave, and we consider these three laws to define what we call *Newtonian physics* today.

Types of Forces and Free-Body Diagrams

Force, F: Vector quantity that can affect the motion of an object or system. Most physicists recognize four fundamental forces in the universe (or a combination of these): **gravitational** forces (F_G), **electromagnetic** forces (F_E), **strong nuclear** forces, and **weak nuclear** forces.

Normal force, F_N or N: Created by the interaction between molecules of two surfaces in contact with each other, is always outward from one surface on the second surface and perpendicular to the surface of contact.

Gravitational force, F_G: Describes the interaction between two objects with mass. The magnitude of the gravitational force on an object with mass m near Earth's surface in the Earth's gravitational field g can be written as shown, where the quantity mg is **weight**:

$$F_G = W = mg$$

The gravitational force is examined in greater detail in Chapter 4.

Tension, F_T or T: The pull on an object by a flexible rope, string, or cord, with the force directed away from the object. If the rope or string is assumed to have no mass and connects two objects, the tension is the same in both directions and on both objects.

Spring force, F_s: May push or pull on an object. It is a variable force that is directly proportional to the displacement $\bar{x}$ of the spring. The spring force and displacement are both vectors, with the negative sign indicating that the displacement is in the opposite direction of the force exerted by the spring. This is a statement of Hooke's law. For an object attached to a spring, the spring exerts the

NGSS HS-PS2-1

33

same magnitude of force on the object as the object exerts on the spring. The spring force is examined in greater detail in Chapter 7.

$$\vec{F}_s = -k\vec{x}$$

Friction force, F_f or f_f: Interaction between two surfaces in contact and is always parallel to the surface of contact and in the opposite direction of the relative motion or attempted motion. The magnitude of the friction force depends upon two factors: (a) the size of the normal force (or essentially how hard the surfaces are pressing on one another), and (b) the **coefficient of friction**, μ, which describes the roughness of the surfaces. These are described by the equation:

$$\mathbf{F}_f = \mu \mathbf{F}_N$$

Friction forces are exerted along the surface of contact between two surfaces, and the direction on each surface is in the opposite direction of motion or attempted motion by the other surface.

When two surfaces are in contact, the friction force is the force of one object on the other—but in the opposite direction. It should be noted that the coefficient of **static friction, μ_s,** between two objects at rest relative to each other is generally larger than the coefficient of **kinetic friction, μ_k,** when the objects are moving relative to each other.

Systems: A system can be a single object or set of objects defined by the observer. It is important to define the system prior to examining forces, since forces *inside the defined system do not change the motion of the system*. **External forces** are those outside the defined system that may change the motion or the properties of the system. For example, we might examine the Moon in nearly circular motion around the Earth. We might define the system as just the Moon—so we would set it up as a circular motion problem. Or, we might set up the problem for the Earth-Moon system as it moves almost like a single object in nearly circular motion around the Sun.

In problems where there is a force internal to the system of objects that you do not need to identify, you may want to treat the entire system as one more complex object. Then identify the external forces in a free-body diagram (see below) and solve for those forces. Two common examples in physics are: (1) tension forces that connect two or more objects that you may want to treat as internal forces, or (2) normal forces between connected objects that are internal to the system. Examples of these show up later in examples that point out the "system approach" to the solution.

Free-Body Diagrams: A free-body diagram is a drawing the identifies all the external forces being exerted on the object or system. Some guidelines for free-body diagrams (sometimes called force diagrams):

- represent the object with a dot so it is free of rotation

- draw only the forces exerted *on* the object.

- draw forces in contact with the dot representing the object

- label each force and show it as a vector in the direction of the force

- direction of the force, pointing away from the object

- include only forces acting *on* the object under consideration.

EXAMPLE—Free-Body Diagram

This is an example of the free-body diagram for an object on a ramp, showing a friction force, a normal force, and the weight or gravitational force. Lengths of arrows indicating magnitudes of forces are generally not required unless requested.

It is important to note that we cannot tell by the free-body diagram whether the object is stationary or moving at constant speed or accelerating. *A component of the gravitational force is directed down the ramp to either balance the friction force or cause the object to accelerate down the ramp.*

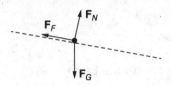

**EXERCISE
3.1**

1. You are pushing downward on a table top with a force of 30 N. How much force is the table top exerting on you—and in what direction?

   A. No force from the tabletop
   B. 30 N normal force upward
   C. 30 N normal force downward
   D. Cannot be determined without knowing the area of contact with the surface

2. When you are walking across a level floor, you need a force pushing forward on your feet to accelerate you forward. What provides the force to push you forward?

   A. friction from floor forward only
   B. friction from floor forward and a normal force from your foot backward
   C. friction force from your foot backward
   D. friction and a normal force forward—both from the floor

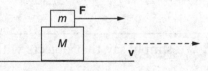

3. In the above situation, what are the forces being exerted on box *M* if both boxes are moving to the right on a smooth floor?

   A. The weight of box *m* downward and the normal force of the floor upward
   B. The pulling force, **F**, to the right, the normal force of the floor upward, and a friction force from the floor to the left
   C. The pulling force, **F**, to the right, the weight of box *m* downward, and the normal force upward from the floor
   D. A normal force from box *m* downward, a normal force of the floor upward, gravitational force on box *M* downward, and a friction force from box *m* to the right

4. In the diagram in Question 3, what are the forces on box *m*?

   A. The weight of box *m* downward and the normal force of the floor upward

   B. The pulling force, **F**, to the right, the normal force of the floor upward, and a friction force from box *M* to the left

   C. The pulling force, **F**, to the right and the weight of box *m* downward and the normal force upward from box *M* and the friction force from box *M* to the left

   D. A normal force from box *m* downward, a normal force of the floor upward, gravitational force on *M* downward, and a friction force from box *m* to the right

5. In the diagram in Question 3, what is the reaction force to the gravitation force of the Earth on box *M*?

   A. the normal force of box *M* on box *m*

   B. the normal force of box *m* on box *M*

   C. the gravitational force of box *M* on the Earth

   D. the weight of box *m* on box *M*

Newton's Laws

Newton's Laws of Motion describe the effects on objects or systems when external forces are exerted on them.

- **Newton's First Law**: An object at rest tends to stay at rest or an object in motion tends to stay in constant motion (i.e., constant speed and direction) unless acted upon by an outside net force. This is called **inertia**.

- **Newton's Second Law**: The acceleration of an object in a given direction is directly proportional to the net force acting upon it in that direction and inversely proportional to the object's mass. This can be stated as an equation:

$$\vec{a}_x = \frac{\Sigma \vec{F}_x}{m} \quad \text{or} \quad \Sigma \vec{F}_x = m\vec{a}_x$$

- **Newton's Third Law**: For every action there is an equal and opposite reaction, with the action and reaction forces acting on different objects.

> Inertia is not a force. It is a property of matter. We simply say that objects with greater mass have more inertia.

EXAMPLE (Advanced)—Gravitational and Normal Forces

In a class experiment, students set up an air track that allows a 0.5-kg cart to "ride" practically frictionless along the track. A motion sensor is set up at one end to determine the acceleration of the cart. The track is tilted at a 30° angle, and the cart is attached by a thin string over a pulley to a 600-g mass hanging over the pulley. What is the predicted acceleration of the cart?

Solution:

(a) Start with a diagram of the setup.

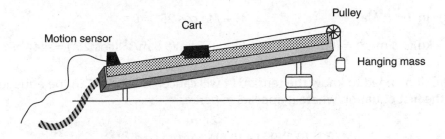

(b) Since there are two moving objects—the cart and the falling mass—construct two free-body diagrams, each with its own frame of reference. Since they are connected, however, they will have the same acceleration.

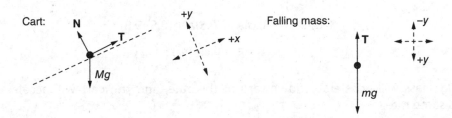

(Note: Within the same problem, don't give two different quantities the same symbol unless you know they are equal. In this case, the massless connecting string has the same tension in both directions, so we've used the same symbol, T, for tension on both diagrams.)

(c) The only force that does not lie along the frame of reference set for each object is the gravitational force on the cart, so that is the only force that needs to be resolved into components.

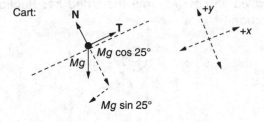

(d) Write Newton's Second Law for the hanging mass:

$$\Sigma \mathbf{F}_y = m\mathbf{a}_y$$

$$mg - \mathbf{T} = m\mathbf{a}$$

$$(0.600 \text{ kg})(9.8 \text{ m/s}^2) - \mathbf{T} = (0.600 \text{ kg})\mathbf{a}$$

(Note: We know the acceleration of both objects must be the same, so we're making the +y direction "down" for the hanging mass so that the acceleration is positive for both objects when the cart slides up the ramp. We're also using the same symbol (**a**) for both objects.)

(e) Write the two Newton's laws for the vectors in two directions (x and y) on the cart.

$$\Sigma F_x = Ma_x \qquad\qquad \Sigma F_y = Ma_y = 0$$

$$T - Mg \sin 25° = Ma_x \qquad\qquad N - Mg \cos 25° = 0$$

$$T - (0.5 \text{ kg})(9.8 \text{ m/s}^2) = (0.5 \text{ kg})a \qquad\qquad N = (0.5 \text{ kg})(9.8 \text{ m/s}^2)(\cos 25°) = 4.44 \text{ N}$$

(f) Since we don't need to know the tension (it will cancel), solve for **T** in the equation in part d and in the first equation where it appears in part e.

$$T = (0.6)(9.8) - (0.6)a \quad \text{(from part d)}$$

$$T = (0.5)a + (0.5)(9.8) \quad \text{(from part e)}$$

(g) Set the above equal to each other and solve the two equations for acceleration **a**:

$$(0.6)(9.8) - (0.6)a = (0.5)a + (0.5)(9.8)$$

$$a = 0.89 \text{ m/s}^2$$

The hanging mass will accelerate downward at this rate, and the cart will accelerate up the track at this same rate.

EXAMPLE—Tension and Gravitational Forces—System Approach

A 1.5-kg block and a 2.5-kg block are suspended by a connecting string over a pulley. Calculate the acceleration of the blocks when the system is released and the instantaneous speed of the masses after they have moved 15 cm. (Assume the string has negligible mass and the pulley has negligible mass and friction.)

Solution:

In analyzing the situation, keep several things in mind: (a) the acceleration will be equal magnitude for both masses but in the opposite direction; (b) since the string has no mass, the tension will be the same throughout; (c) assuming the string doesn't stretch, the masses will move the same distance; and (d) friction will have a negligible effect.

Construct separate free-body diagrams for the two masses with a frame of reference for each one that makes the positive acceleration the same for the two (i.e., the upward acceleration of the smaller mass is positive when the downward acceleration of the larger mass is also positive, since they move together). To make our work simpler, we'll use the diagram to analyze the forces.

Write Newton's Second Law for each of the masses.

Small mass:

$\Sigma F_y = m\mathbf{a}$

$T - mg = m\mathbf{a}$

Large mass:

$\Sigma F_y = m\mathbf{a}_y$

$Mg - T = M\mathbf{a}$

Combine equations and solve for **a**:

Small mass: $T = mg + m\mathbf{a}$

Large mass: $T = Mg - M\mathbf{a}$

$mg + m\mathbf{a} = Mg - M\mathbf{a}$

$(1.5 \text{ kg})(9.8 \text{ m/s}^2) + (1.5 \text{ kg})\mathbf{a} = (2.5 \text{ kg})(9.8 \text{ m/s}^2) - (2.5 \text{ kg})\mathbf{a}$

$4\mathbf{a} = 9.8 \qquad \mathbf{a} = 2.5 \text{ m/s}^2$

Alternate Solution (System Approach):

In a case like this one where we know the masses move as a system, we can take a *system approach* to the solution—with the system consisting of the two masses and the string. We only examine forces external to the system, so the tension (which is internal to the system) cancels and has no effect on the acceleration. We rewrite one Newton's second law for the system (which has a total mass of 4.0 kg) and solve for the acceleration:

$$\Sigma F_{system} = m_{system}\mathbf{a}_{system}$$
$$Mg - mg = (M + m)\mathbf{a}$$
$$(2.5)g - (1.5)g = (4.0)\mathbf{a}$$
$$\mathbf{a} = 2.5 \text{ m/s}^2$$

Use kinematics to find the final speed of the system, with the large mass moving downward and the small mass moving upward.

$$\mathbf{v}_f^2 = \mathbf{v}_0^2 + 2\mathbf{a}s$$
$$\mathbf{v}_f^2 = (0) + (1)(2.45 \text{ m/s}^2)(0.15 \text{ m})$$
$$\mathbf{v}_f = \sqrt{(2.45)(0.15)} = 0.61 \text{ m/s}$$

EXAMPLE—Tension and Gravitational Forces

An emergency worker must lower a bucket of supplies to a coworker in the bottom of a well as quickly as possible without breaking the rope holding the bucket or damaging the contents of the bucket when it lands at the bottom. The bucket of supplies weighs 600 N, and the rope will only withstand a maximum tension of 500 N. What can be done?

Solution:

Obviously, if the rope is released completely, the bucket will fall and damage the contents or hurt the worker below. Start with a free-body diagram of the bucket to determine just how quickly it can be lowered.

Write Newton's second law equation to solve for maximum acceleration as the bucket is lowered:

$$\Sigma \mathbf{F}_y = m\mathbf{a}$$

$$\mathbf{T}_{rope} - \mathbf{W}_{bucket} = m_{bucket}\mathbf{a}$$

$$500 \text{ N} - 600 \text{ N} = \left(\frac{600 \text{ N}}{9.8 \text{ m/s}^2}\right)\mathbf{a}$$

$$\mathbf{a} = -1.6 \text{ m/s}^2$$

We can see from the solution that the "physics works" as long as the bucket is accelerated downward at 1.6 m/s². We can also see that the bucket cannot be lowered at constant velocity, or the rope would break. The rope would obviously also break if the bucket is lifted.

The life skill to know here is that if a rope might break under a load, accelerating downward might keep it from breaking!

EXAMPLE—Tension and Gravitational Forces

Now the same rope as in the previous example has to be used to bring the coworker back to the top—but the coworker weighs 800 N. If he ties onto the rope and someone tries to pull him up from the top, the rope will break. What can be done?

Solution:

The workers know that the rope won't hold the stranded worker's weight. However, if the worker at the top throws the rope over a pulley or around a tree—anything that will let the rope slip around it—and throws the other end down to the worker, the stranded worker can pull up alone by tying onto one end and pulling on the other end. Both ends of the rope are now attached to the worker. Examine a free-body diagram:

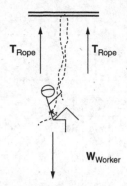

T_{Rope} T_{Rope}

W_{Worker}

As the worker pulls down on the loose end of the rope, there is a tension in the opposite direction that pulls upward. Let's assume the worker pulls at constant speed.

Using this method, the worker can easily be supported by the rope, since the tension is half the worker's weight. Notice that there are two forces upward and one force downward exerted on the worker. The worker can pull downward with 400 N and lift himself or herself at constant speed.

$$\Sigma F_y = 0$$

$$T + T - W = 0$$

$$2T = 800 \text{ N}$$

$$T = 400 \text{ N}$$

This is another "physics life skill." Using a single fixed pulley, you can lift yourself with a force equal to half your weight—but if someone else grabs the rope to lift you, they will have to pull with your entire weight, since there is only one rope attached to you.

EXAMPLE—Normal and Friction and Gravitational Forces

A small 0.1-kg block sits on top of a larger block on a smooth floor. (a) Find the maximum force, **F**, that will accelerate the two blocks together without the blocks slipping. The coefficient of friction between the blocks is 0.4, and the large block has twice the mass of the small block. (b) Find the acceleration of the blocks.

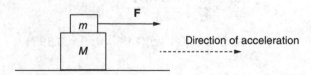

Solution:

First, draw a free-body diagram for each block to determine direction of friction force on each block. Start with the upper block:

N_B

m **F**

F_F

mg

Now the lower block:

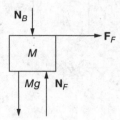

We can see that without a friction force at the floor, there is an unbalanced force in the x-direction on the lower block—so it must accelerate. Thus, the accelerations of the two blocks must be the same if there is no slippage between the blocks.

In the y-direction, the normal force (N_B) of the lower block upward on the upper block is equal to the reaction force (N_B) of the upper block downward on the lower block.

The friction force (F_f) must be to the right on the lower block because that is the only horizontal force acting to move that block to the right. Therefore, the reaction force on the upper block is the friction force (F_f), which must be to the left. The two friction forces are equal and opposite in direction.

Write the Newton's law equations in each direction for each block using the same acceleration (**a**) for both blocks:

	Upper block	Lower block
x-direction:	$\mathbf{F} - \mathbf{F}_f = m\mathbf{a}$	$\mathbf{F}_f = M\mathbf{a}$
y-direction:	$\mathbf{N}_B = mg$	$\mathbf{N}_B + Mg = \mathbf{N}_f$

Substituting from the above equations into the coefficient of friction equation for the friction force between the two blocks:

$$\mathbf{F}_f = \mu\mathbf{N}_B = \mu mg = M\mathbf{a} = 2m\mathbf{a}$$

$$\mathbf{F} = \mathbf{F}_f + m\mathbf{a} = M\mathbf{a} + m\mathbf{a} = 3m\mathbf{a}$$

By definition: $\qquad \mathbf{F}_f = \mu\mathbf{N}_B$

So: $\qquad 2m\mathbf{a} = \mu mg$

$$\mathbf{a} = \frac{\mu g}{2} = \frac{(0.4)(9.8 \text{ m/s}^2)}{2} = 1.96 \text{ m/s}^2$$

$$\mathbf{F} = 3m\mathbf{a} = (3)(0.1 \text{ kg})(1.96 \text{ m/s}^2) = 0.59 \text{ N}$$

Notice that the acceleration was found independently of the mass of the blocks, so this acceleration would be true for any two blocks where the bottom block has twice the mass of the other and the coefficient of friction is 0.4.

EXAMPLE—Spring and Normal and Gravitational Forces

A student pulls on the spring scale shown in the diagram, which accelerates a 2.0-kg block in the positive x-direction. A motion sensor records the motion and creates the plot below. Once the block starts moving, the spring scale reads 2.6 N. From the plot, determine: (a) the average acceleration of the block after it starts moving, (b) the average force exerted on the block after it starts moving, and (c) the coefficient of kinetic friction between the block and table.

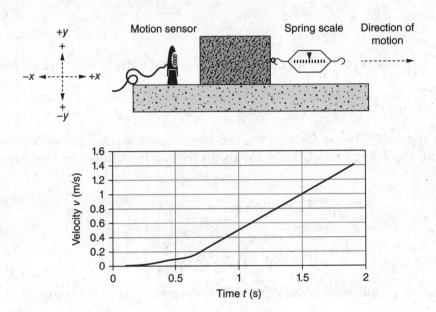

Solution:

In determining the acceleration, we'll ignore the curve during the first 0.7 s, since this is when the force is overcoming static friction to start the block moving.

(a) Using the consistent part of the graph from 0.7 to about 1.8 s, determine the slope to find acceleration.

$$\text{slope} = \frac{\Delta y}{\Delta x} = \frac{(1.4 - 0.1)\text{ m/s}}{(1.8 - 0.7)\text{ s}} = 1.2 \text{ m/s}^2$$

(b) Use the acceleration and the mass of the box to find the average force during the same time period, 0.7 to 1.8 s.

$$\Sigma \vec{F} = m\vec{a}$$

$$\mathbf{F}_{pulling} - \mathbf{F}_{fr} = ma$$

$$2.6 \text{ N} - \mathbf{F}_{fr} = (2.0 \text{ kg})(1.2 \text{ m/s}^2)$$

$$\mathbf{F}_{fr} = 0.2 \text{ N}$$

$$\mathbf{F}_{fr} = \mu_k mg = \mu_k (2.0 \text{ kg})(9.8 \text{ m/s}^2)$$

$$\mu_k = 0.010$$

EXAMPLE—System Approach to Solution

Consider the pulley apparatus below with a block on a table connected by a rope to a hanging block.

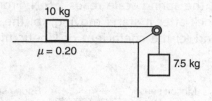

(a) Find the acceleration of the two-block system. (b) What is the tension in the rope?

Solution:

(a) (System approach) The net force on the two-block system is equal to the downward force of gravity on the 7.5-kg block (73.5 N) minus the friction force between the upper block and the table:

$$\mathbf{F}_{net} = mg - \mathbf{F}_f$$

$$\mathbf{F}_f = \mu Mg = 0.2(10)(9.8) = 19.6 \text{ N}$$

$$\mathbf{F}_{net} = 73.5 \text{ N} - 19.6 \text{ N} = 53.9 \text{ N}$$

Then find the acceleration of the system using the total mass of the system:

$$\mathbf{a} = \frac{\mathbf{F}_{net}}{m + M} = \frac{53.9 \text{ N}}{17.5 \text{ kg}} = 3.08 \text{ m/s}^2.$$

(b) To find the tension, consider only the 10-kg block on the table. The tension minus friction force is equal to the net force on the 10-kg block. The friction force was previously calculated as 19.6 N and the acceleration as 3.08 m/s².

$$\Sigma \mathbf{F} = m\mathbf{a}$$

$$\mathbf{T} - 19.6 \text{ N} = (10 \text{ kg})(3.08 \text{ m/s}^2) = 30.8 \text{ N}$$

$$\mathbf{T} = 50.4 \text{ N}$$

EXERCISE
3.2

1. If an object with a mass of 3 kg has two forces exerted on it—a force of 10.5 N to the east and a force of 6.4 N to the west—what will be the acceleration of the object as a result?

   A. 1.4 m/s/s toward the east
   B. 5.6 m/s/s toward the east
   C. 2.1 m/s/s toward the east
   D. 2.1 m/s/s toward the west

2. The graph below plots data for the acceleration of an object as a function of the net force exerted on that object. Use the graph to determine the mass of the object.

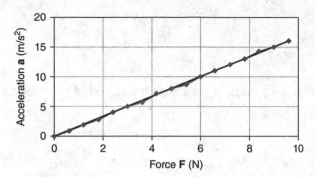

A. 0.2 kg
B. 0.6 kg
C. 1.7 kg
D. 2.4 kg

3. An automobile that is towing a trailer accelerates on a level road. The force that the automobile exerts on the trailer is:

A. equal to the force that trailer exerts on the road
B. greater than the force the trailer exerts on the automobile
C. equal to the force the trailer exerts on the automobile
D. equal to the force the road exerts on the trailer

4. A 2.0-kg cart is pulled across a horizontal surface. If the horizontal pulling force on the cart is 12 N when $a = 5$ m/s², what is the friction force?

A. 0
B. 0.5 N
C. 1.0 N
D. 2.0 N

5. (Advanced) A box is held stationary on a ramp by a string connected to the wall. The forces on the box are labeled in the diagram below. Which statement is true?

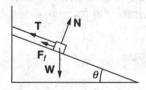

A. $T = -F_f$
B. $T = W \cos\theta + F_f$
C. $W \sin \theta = T + F_f$
D. $T + F_f = W$

6. An elevator weighing 10,000 N is supported by a steel cable. What is the tension in the cable when the elevator is being accelerated upward at a rate of 3.0 m/s²?

A. 13,000 N
B. 10,000 N
C. 7,000 N
D. 4,000 N

7. An object of mass 15 kg has a net force of 10 N exerted on it. What will be the acceleration of the object, in m/s^2?

   A. 150
   B. 15
   C. 1.5
   D. 0.67

8. A force of 10 N, when applied for 5 s to a 2-kg mass, will change the speed of the mass:

   A. from rest to 12 m/s
   B. from 10 m/s to 25 m/s
   C. from 20 m/s to 50 m/s
   D. from 20 m/s to 45 m/s

9. Two perpendicular forces, one of 30 N directed due north and the second, 40 N directed due east, are exerted simultaneously on one object. The object has a mass of 35 kg. Determine the magnitude of the resultant acceleration of the object (in m/s^2).

   A. 155
   B. 3.5
   C. 2.1
   D. 1.4

10. A massless string connects three blocks as shown below. A force of 12 N acts on the system. Assuming no friction with the surface, what is the acceleration of the blocks and the tension in the rope attached to the 1-kg block?

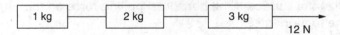

Acceleration	Tension
A. 4 m/s^2	6 N
B. 2 m/s^2	2 N
C. 2 m/s^2	12 N
D. 12 m/s^2	6 N

11. A person is standing on a scale in an elevator that is accelerating upward. Compare the reading on the scale to the person's true weight, which is the gravitational force exerted by Earth on the person.

   A. The scale reading will always be the same as the person's weight.
   B. The scale reading will be greater than the person's weight.
   C. The scale reading will be less than the person's weight.
   D. The scale will always read zero when the elevator is moving.

12. A force of 100 N is applied to a block on a smooth surface, causing the block to accelerate horizontally with a magnitude of 2.5 m/s/s. Two additional identical blocks are now stacked on top of the original block. What force would be required to make the stack of blocks accelerate horizontally at the same rate?

   A. 300 N, since the mass is three times as much
   B. 200 N, since the mass has been doubled
   C. 100 N, since the same block is in contact with the surface
   D. 40 N, since a = F/m

13. In a lab simulation, a force of 100 N is applied forward to a box with a mass of 20 kg, and when the simulation is run, the box accelerates forward at 4 m/s/s. Explain why the acceleration is 4 m/s/s instead of 5 m/s/s.

    A. There must be friction force of 20 N pushing backward on the cart at the same time.
    B. There must be friction force of 5 N pushing backward on the cart at the same time.
    C. There must be an upward force of the floor on the cart that is decreasing the acceleration rate.
    D. The gravitational force of the Earth on the box is decreasing the acceleration.

14. The following forces are exerted on a 10-kg object on a level surface that has negligible friction: 10 N <u>to the east</u>, 20 N <u>to the east</u>, 30 N <u>to the north</u>, and 30 N <u>to the west</u>. Will the object accelerate, and if so, what will be the acceleration?

    A. The object will accelerate at 3 m/s/s to the west.
    B. The object will accelerate at 0.3 m/s/s to the north.
    C. The object will accelerate at 3 m/s/s to the north.
    D. The object will not accelerate due to the force exerted on it.

Equilibrium

Equilibrium is a situation where the net force on an object or system is zero; that is, the forces are balanced. In equilibrium, an object has an acceleration of zero (see Newton's Second Law). An object or system in **static equilibrium** has no net force on it and is not moving. An object or system in **dynamic equilibrium** has no net force on it and is moving at constant velocity.

EXAMPLE—Equilibrium with Components (Qualitative)

In this case, an object is sitting stationary on a tilted ramp. What does this diagram reveal about the forces on the object?

Solution:

It's a good idea to work this problem from an angle by setting up the x- and y-axes for reference.

Since the object is not accelerating, the net force in each dimension (x and y) must be zero.

Therefore, we can deduce that the friction force upward on the ramp is equal to the component of gravitational force down the ramp. Also, the normal force in the positive y-direction must be equal to the component of the gravitational force that is directed in the negative y-direction.

EXAMPLE—Tension and Gravitational Forces
(Advanced with Components)

Two cables are attached to the roof, and a piano is held up by the cables. If the piano weighs 350 lbs., find the tension in each of the cables.

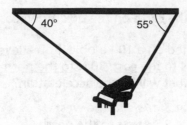

Solution:

This is a situation of static equilibrium, where the net force in each dimension is zero. We'll consider the piano as the system and set up a free-body diagram (with components) to examine the forces on the piano.

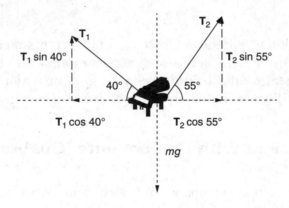

x-direction: $\Sigma \mathbf{F}_x = 0$

$\mathbf{T}_2 \cos 55° - \mathbf{T}_1 \cos 40° = 0$

y-direction: $\Sigma \mathbf{F}_x = 0$

$\mathbf{T}_2 \sin 55° + \mathbf{T}_1 \sin 40° - mg = 0$

Convert the piano's weight:

$$(350 \text{ lb})\left(\frac{1 \text{ kg}}{2.2 \text{ lb}}\right) = 159 \text{ kg}$$

$$\mathbf{W} = mg = (159 \text{ kg})(9.8 \text{ m/s}^2) = 1{,}560 \text{ N}$$

One way to solve the equations above is by substitution. Solve for T_1 in the first equation and substitute that into the second equation. Then solve for T_2.

x-direction: $0.77 \text{ T}_1 = 0.57 \text{ T}_2$

$\text{T}_1 = 0.74 \text{ T}_2$

y-direction: $0.64\,T_1 + 0.82\,T_2 - 1560 = 0$

$0.64(0.74\,T_2) + 0.82\,T_2 = 1560$

$T_2 = \dfrac{1560}{1.29} = 1210\ \text{N}$

$T_1 = (0.74)(930) = 690\ \text{N}$

> It makes sense that the cord that is more vertical is holding more of the piano's weight, so $T_2 > T_1$.

EXERCISE
3.3

1. A horizontal 100-N force is exerted on an 8.0-kg chair to push it at constant speed across a level floor. Which statement has to be true?

   A. The chair accelerates at 12.5 m/s².
   B. The applied force is greater than the friction force.
   C. The friction force is 100 N.
   D. There is no friction between the floor and chair.

2. A 2.0-kg cart is pulled across a horizontal surface. If the horizontal pulling force on the cart is 12 N and the cart does not accelerate, what is the friction force?

   A. 0
   B. 0.5 N
   C. 1.0 N
   D. 12.0 N

3. A box with a weight of 500 N is at rest on a smooth, level surface. The box exerts a force downward on the surface. What is the force exerted by the surface upward on the box?

   A. 500 N upward due to Newton's third law
   B. Zero, since the box is not accelerating either upward or downward
   C. 250 N upward, since the surface is only half of the two surfaces
   D. Zero due to Newton's first law

4. (Advanced) A traffic light that weighs 100 N is supported by two ropes, as shown in the figure below. What is the tension in each rope?

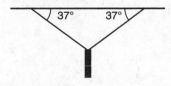

   A. Less than 100 N but more than 50 N
   B. Equal to 50 N
   C. More than 100 N
   D. Equal to 100 N

5. The piano in this diagram is in equilibrium. What must be true of $T_1 \cos 40°$?

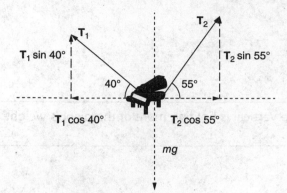

A. It must be equal to $T_1 \sin 40°$.
B. It must be equal to $T_2 \cos 55°$.
C. It must be equal to $T_2 \cos 40°$.
D. It must be equal to $T_2 \sin 55°$.

Gravitation, Circular Motion, and Projectile Motion

·4·

- Gravitational Force
- Gravitational Field and Gravitational Mass
- Centripetal Force and Uniform Circular Motion
- Projectile Motion

Gravitational force, the force that seems to dominate us at the human scale, is actually the weakest of the fundamental forces in the universe. We will examine the gravitational force and weight, the gravitational force providing the centripetal force for objects in circular orbits, the factors that affect circular motion, and other forces that provide the centripetal force for objects in circular motion.

Gravitational Force

Gravity or Gravitational Force: The attraction between objects with mass, measured in newtons (N). Newton derived his *Universal Law of Gravitation*, which describes the fact that "every object in the universe has a gravitational attraction for every other object in the universe." Those gravitational forces between two objects are equal in magnitude and opposite in direction. For any two objects with masses M and m and a distance r between their centers of mass:

$$\mathbf{F}_{grav} = -\frac{GMm}{r^2}$$

In this case, the Earth attracts the satellite with a force **F** that is exactly equal in magnitude and opposite in direction to the force the satellite exerts on the Earth.

The negative sign is a way of indicating that the force is attractive, and G is the universal gravitational constant:

$$G = 6.67 \times 10^{-11} \frac{N \times m^2}{kg^2}$$

NGSS HS-PS2-4

51

1. Assume the Sun's mass is about 300,000 times the mass of the Earth and radius is about 100 times Earth's radius. Compare the gravitational force on an object near Sun's surface to the gravitational force on the same object near Earth's surface.

   A. 10,000 times
   B. 3,000 times
   C. 300 times
   D. 30 times

2. Three objects, each with a mass of 2 kg, are positioned along a line with coordinates as shown below. Which object has the greatest net gravitational force on it due to the other two?

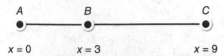

   A. The net force on *A* is the greatest.
   B. The net force on *B* is the greatest, and the net force on *A* and on *C* are the same.
   C. The net force on *C* is the greatest.
   D. The net force on *A* and on *C* are the same and are greater than the net force on *B*.

3. A satellite is in a circular orbit around the Earth at an altitude of three Earth radii above the surface. How does the gravitational force of Earth on the satellite at this altitude compare to the gravitational force of Earth on the satellite when it is sitting on the surface prior to launch?

   A. None; the satellite has no gravitational force on it when it is moving in orbit.
   B. The gravitational force is ½ as great in orbit as on the surface of Earth.
   C. The gravitational force is ¼ as great in orbit as on the surface of Earth.
   D. The gravitational force is ¹⁄₁₆ as great in orbit as on the surface of Earth.

4. Two binary stars have masses of *M* and 2*M*, respectively. Find the ratio of gravitational force the smaller star exerts on the larger star to the gravitational force the larger star exerts on the smaller star.

   A. 1:1
   B. 1:2
   C. 2:1
   D. 4:1

5. The Moon's mass is about one-sixth of the Earth's mass. Compared to the gravitational force the Earth exerts on the Moon, the gravitational force the Moon exerts on the Earth:

   A. is one-sixth as much
   B. is one-half as much
   C. is the same
   D. is twice as much

Gravitational Field and Gravitational Mass

Gravitational Field: Region of influence that surrounds anything with mass; another object with mass interacts with the field, resulting in a gravitational force between the two objects. The gravitational field near the Earth's surface is 9.8 N/kg and gets weaker farther away from Earth.

Gravitational Acceleration: Acceleration of an object with mass in the field of another object with mass; for objects near the Earth's surface, this is called g and is equal to 9.8 m/s^2.

Weight: The gravitational force (in Newtons) on an object near the Earth's surface, $W = mg$.

Weightlessness: An astronaut aboard a space shuttle will experience what we call *weightlessness* only because the astronaut, the shuttle, and everything and everyone on board are all in free fall; i.e., they are accelerating toward Earth since the gravitational force is the only force on them. They don't hit Earth because the tangential velocity of the shuttle is large enough.

Gravitational Mass: The mass of an object as determined by its weight in a gravitational field. For example, an object that weighs 120 newtons has a mass determined by the equation: $W = mg$. Its mass is $m = W/g = 120$ N/9.8 m/s^2, which is 12.2 kg.

Inertial Mass: The mass of an object as determined by its motion—not by its weight. Inertial mass could be determined, for example, in space—where gravity is absent or unknown. One way to do this would be to attach the object to a spring and use the spring equation below by determining the period or frequency of oscillation of the spring. An object with larger mass causes the spring to oscillate more slowly (i.e., have a larger period of oscillation).

$$T = 2\pi\sqrt{\frac{m}{k}}$$

EXAMPLE—Calculate *g*

Find: (a) an expression for g in terms of the Earth's mass and (b) the effective value for g at an altitude of 400 mi above the Earth's surface—about where the International Space Station orbits.

Solution:

(a) Set the expression for weight of an object, mg, equal to the formula for gravitational force and solve for g.

$$mg = \frac{GM_{Earth}m}{R^2}$$

$$g = \frac{GM_{Earth}}{R^2}$$

(b) Use the formula, with orbital radius equal to radius of Earth, R plus orbital height, h:

$$g = \frac{(6.67 \times 10^{-11}\ \text{N-m}^2/\text{kg}^2)(5.97 \times 10^{24}\ \text{kg})}{(6.37 \times 10^6\ \text{m} + 6.44 \times 10^5\ \text{m})^2} = 8.1\ \text{m/s}^2$$

---

1. The gravitational acceleration, g, on the surface of a planet varies as a:

   A. direct proportion to the mass of the planet and inverse proportion to its radius squared
   B. direct proportion to the radius of the planet and inverse proportion to its mass
   C. direct proportion to the square root of mass and inverse proportion to square root of radius squared
   D. direct proportion to the square root of radius and inverse proportion to square root of mass

2. The weight of an object on the surface of the Earth is 40 N. The radius of Earth is approximately 6,400 m. What would be the weight of the object if it is located 6,400 m above Earth's surface?

   A. 40 N
   B. 20 N
   C. 10 N
   D. 5 N

3. An object with a mass of 60 kg on the surface of Earth is taken to the Moon, where the gravitational field value, g, is $\frac{1}{6}$ of that on Earth. What is the mass of the object on the surface of the Moon?

   A. 60 kg
   B. 40 kg
   C. 30 kg
   D. 10 kg

4. The theoretical value of g on Earth's surface at the equator, based upon the gravitational force between the Earth and any object on the surface of Earth, is considered to be 9.8 m/s². However, the effective value of g at the equator is always slightly less, primarily due to:

   A. The normal force of Earth's surface
   B. The spin of Earth on its axis
   C. A lack of knowledge about the exact mass of Earth
   D. A lack of knowledge about the exact radius of Earth

5. Which of the following methods would be reasonable to determine the inertial mass of an object?

   A. Find the normal force the object exerts on a surface.
   B. Attach the mass to a pendulum and find the period of oscillation.
   C. Attach the mass to a horizontal spring and find the period of oscillation.
   D. Find the buoyant force when the object is submerged in water.

Centripetal Force and Uniform Circular Motion

Centripetal Acceleration: Acceleration toward the center of circular motion. An object in motion in a circle is constantly accelerating since it is constantly changing the direction of its velocity—even though it may be moving at constant speed.

$$\mathbf{a}_c = \frac{\mathbf{v}^2}{R}$$

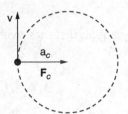

Centripetal Force: Force exerted on an object moving in a circle that is directed toward the center of the motion; force that changes the direction of tangential motion, directing the motion so that the object maintains a constant distance (radius) from center of motion. In the formula, m is the mass of the object moving in the circle, $\mathbf{v}$ is the tangential velocity of the object, and R is the radius of the circle.

$$\mathbf{F}_c = \frac{m\mathbf{v}^2}{R}$$

> In the centripetal force formula, it's important to note that the mass is that of the object moving in the circle. The centripetal force and centripetal acceleration are both toward the center.

EXAMPLE—Motion in a Horizontal Circle

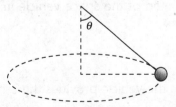

A ball with a mass of 0.20 kg is held by a string and whirled in a horizontal circle in a radius of 0.60 m. The length of the string is 0.80 m.

(a) Calculate the angle θ.

(b) The ball makes 10 complete revolutions in 8.0 s. Calculate the period of revolution of the ball as it moves in the circle.

(c) Calculate the speed of the ball as it moves in the circle.

(d) Calculate the tension in the string when the ball is moving at this speed.

Solution:

(a) Use the lengths given for the string and the radius to calculate the angle:

$$\sin\theta = \frac{R}{L} = \frac{0.6\text{ m}}{0.8\text{ m}} = 0.75$$

$$\theta = 49°$$

(b) $T = \dfrac{8.0\text{ s}}{10\text{ rev}} = 0.8\text{ s}$

(c) $v = \dfrac{2\pi R}{T}$ Think of this as "velocity equals distance over time."

$$v = \frac{2\pi(0.8\sin 49°)}{0.8\text{ s}} = 4.7\text{ m/s}$$

(d) For the ball to move in a constant horizontal circle, as shown, the net vertical force must equal zero, and the net horizontal force is the centripetal force.

$$\Sigma F_y = 0$$

$$T\cos\theta = mg$$

$$T(\cos 49°) = (0.20\text{ kg})(9.8\text{ m/s}^2)$$

$$T = 3.0\text{ N}$$

> "Centripetal force" is never a correct label on a free-body diagram. *Centripetal* only describes the direction of the force. You must always ask, "What is the force providing the centripetal force in this situation?"

EXAMPLE—Circular Orbit

A space vehicle orbits a certain planet at an orbital radius of 6.0×10^6 meters and orbital period of 6,000 s. What is the average speed of the space vehicle in its orbit? (b) What is the mass of the planet?

Solution:

The planet's gravitational force on the vehicle provides the centripetal force for the orbit.

(a) $v = \dfrac{2\pi R}{T} = \dfrac{2\pi(6\times10^6\text{ m})}{6,000\text{ s}} = 6,383\text{ m/s}$

(b) $F_G = F_C$

$$\frac{GM_E m_s}{R^2} = \frac{m_s v^2}{R}$$

$$\frac{GM_E \cancel{m_s}}{R^{\cancel{2}}} = \frac{\cancel{m_s} v^2}{\cancel{R}}$$

$$M = \frac{v^2 R}{G} = \frac{(6,383\text{ m/s})^2(6\times10^6\text{ m})}{6.67\times10^{-11}\text{ N}\times\text{m}^2/\text{kg}^2} = 3.67\times10^{24}\text{ kg}$$

EXAMPLE—Satellite in a Circular Orbit

Consider a small satellite with a mass of 10 kg in orbit around the Earth at a height of 500 km above the Earth's surface. Calculate the velocity of the satellite in this orbit.

Solution:

The gravitational force of Earth on the satellite provides the centripetal force for the orbit (assuming a nearly circular orbit).

$$\mathbf{F}_G = \mathbf{F}_C$$

$$\frac{GM_E m_s}{R^2} = \frac{m_s \mathbf{v}^2}{R}$$

The radius in the orbit, R, is the center-to-center distance between the orbiting body and the body being orbited (central body). In this case, that is the radius of Earth plus the distance of the satellite above Earth. We'll assume the radius of the satellite is too small to be considered in the calculation.

$$R_{orbit} = R_{Earth} + h = 6{,}380 \text{ km} + 500 \text{ km} = 6{,}880 \text{ km}$$

Now substitute values to find $\mathbf{v}$:

$$\frac{GM_E \cancel{m_s}}{R^{\cancel{2}}} = \frac{\cancel{m_s} \mathbf{v}^2}{\cancel{R}}$$

$$\mathbf{v} = \sqrt{\frac{GM}{R}} = \sqrt{\frac{(6.67 \times 10^{-11} \text{ N} \times \text{m}^2/\text{kg}^2)(5.97 \times 10^{24} \text{ kg})}{6.88 \times 10^6 \text{ m}}} = 7{,}608 \text{ m/s}$$

EXAMPLE—Car Turning in a Circular Curve

A car travels around a circular turn of radius 50 meters while maintaining a constant speed of 15 m/s. (a) Draw and clearly label a force diagram showing all forces acting on the car in this problem. (b) Find the minimum value for the coefficient of friction necessary to keep the car on the road. (c) In icy conditions, the coefficient of friction drops to 0.15. Now find the maximum safe speed for making the turn.

Solution:

(a)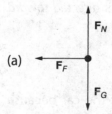

(b) There is no net force in the y-direction, so the weight of the car is equal to the normal force. In the x-direction, the only net force is the friction force, which provides the centripetal force for the car to make the turn.

$$\Sigma F_x = F_f$$

$$\Sigma F_y = 0$$

$$F_f = \mu N = \mu mg = \frac{mv^2}{r}$$

$$\mu = \frac{v^2}{gr} = \frac{(15\,\text{m/s})^2}{(9.8\,\text{m/s}^2)(50\,\text{m})} = 0.46$$

(c) Using the same reasoning, substitute the new value for the coefficient of friction and solve for **v**:

$$\mu = \frac{v^2}{gr}$$

$$v = \sqrt{\mu gr} = \sqrt{(0.15)(9.8\,\text{m/s}^2)(50\,\text{m})} = 8.6\,\text{m/s}$$

EXAMPLE (Advanced)—Banked Curve

A curve of radius 40 m is banked at 30°. Suppose that an ice storm hits, and the curve is effectively frictionless. What is the safe speed with which to take the curve without sliding up or down the bank? (Show a free-body diagram.)

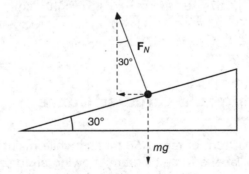

Solution:

Step 1: We look for the force—or force component—that is directed toward the center of the circle. It is this force that provides the centripetal force to keep the car turning on the curve. We can see from the diagram above that it is the horizontal component of the normal force that is directed toward the center.

$$F_N \sin\theta = \frac{mv^2}{R}$$

The vertical component of the normal force must be equal to the weight of the car because there is no net force in the *y*-direction. Thus

$$F_N \cos\theta = mg$$

To solve using these two equations, one method would be to solve one equation for F_N and substitute into the second. Another method would be to solve both equations for F_N and set the two equations equal to each other. We'll use an even different method: divide one equation by the other equation.

$$\frac{F_N \sin\theta}{F_N \cos\theta} = \frac{\frac{mv^2}{R}}{mg}$$

$$\tan\theta = \frac{v^2}{gR}$$

$$v = \sqrt{gR\tan\theta} = \sqrt{(9.8\,m/s^2)(40\,m)(\tan 30°)} = 15\,m/s$$

EXAMPLE—Geosynchronous Satellite

Geosynchronous satellites orbit Earth at an altitude and velocity such that they maintain a constant position relative to the Earth's surface. Find the altitude above the surface for geosynchronous satellites.

Solution:

For a satellite to be geosynchronous and maintain a constant position relative to the surface, the satellite must make one complete orbit in the same time that the Earth rotates once. That is, the satellite's orbital period must be one day.

$$T = 24\,h = 86,400\,s$$

Next, find the velocity of the orbit.

$$v = \frac{2\pi R}{T}$$

Use the concept that the gravitational force provides the centripetal force:

$$F_G = F_C$$

$$\frac{GM_E m_s}{R^2} = \frac{m_s v^2}{R}$$

$$\frac{GM_E m_s}{R^2} = \frac{m_s v^2}{R}$$

$$v = \sqrt{\frac{GM}{R}}$$

Set these two expressions equal to each other because they are both equal to v:

$$\frac{2\pi R}{T} = \sqrt{\frac{GM}{R}}$$

Solve for R:

$$4\pi^2 R^3 = GMT^2$$

$$R = \sqrt[3]{\frac{GMT^2}{4\pi^2}} = \sqrt[3]{\frac{(6.67 \times 10^{-11}\,\text{N}\times\text{m}^2/\text{kg}^2)(5.97 \times 10^{24}\,\text{kg})(86,400\,\text{s})}{4\pi^2}} = 9.55 \times 10^5\,\text{m}$$

Use that value for R to solve for $\mathbf{v}$:

$$\mathbf{v} = \sqrt{\frac{GM}{R}} = \sqrt{\frac{(6.67 \times 10^{-11}\,\text{N}\times\text{m}^2/\text{kg}^2)(5.97 \times 10^{24}\,\text{kg})}{9.55 \times 10^5\,\text{m}}} = 20,420\,\text{m/s}$$

EXAMPLE—Moon in Orbit

If we assume an approximately circular orbit for the Moon around the Earth, determine the average speed of the Moon as it travels in its orbit. (The distance of the Moon from Earth is about 240,000 miles, and the mass of the Moon is about 1/80 the mass of Earth.)

Solution:

If you don't remember the formula to calculate speed of an object in orbit, it's always helpful to remember that the gravitational force of the Earth on the Moon provides the centripetal force for the nearly circular orbit. (The real truth is that the Sun has a significant influence on the Moon- Earth system, locking them in a "dance" around each other as they move around the Sun. However, we'll make estimates here.)

Step 1. Convert units:

$$(240,000\,\text{mi})\left(\frac{1,609\,\text{m}}{1\,\text{mi}}\right) = 3.86 \times 10^8\,\text{m}$$

Step 2. Set gravitational force for centripetal force and find velocity:

$$\mathbf{F}_G = \mathbf{F}_C$$

$$\frac{GM_E m_m}{R^2} = \frac{m_m \mathbf{v}^2}{R}$$

$$\frac{GM_E \cancel{m_m}}{R^{\cancel{2}}} = \frac{\cancel{m_s} \mathbf{v}^2}{R}$$

$$\mathbf{v} = \sqrt{\frac{GM}{R}} = \sqrt{\frac{(6.67 \times 10^{-11}\,\text{N}\times\text{m}^2/\text{kg}^2)(5.97 \times 10^{24}\,\text{kg})}{3.86 \times 10^8\,\text{m}}} = 1,016\,\text{m/s}$$

EXAMPLE—Motion in a Vertical Circle

A string that is 0.8 meter long is used to spin a 2-kg stone in a horizontal circle. If the maximum tension the string can hold is 120 N, calculate the maximum speed of the stone.

(Note: It is impossible to spin the stone in a perfectly horizontal circle.)

Solution:

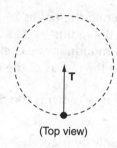

(Top view)

The net force toward the center of the motion provides the centripetal force. Since the weight of the stone is downward toward the floor, it does not figure into the calculation.

$$\Sigma F = \frac{m\mathbf{v}^2}{R}$$

$$T = \frac{m\mathbf{v}^2}{R}$$

$$120\,N = \frac{(2\,kg)\mathbf{v}^2}{0.8\,m}$$

$$\mathbf{v} = \sqrt{\frac{(120)(0.8)}{2}} = 6.93\,m/s$$

EXERCISE 4.3

1. A car goes around a curve of radius *r* at a constant speed *v*. Then it goes around a curve of radius *2r* at speed *2v*. What is the centripetal acceleration of the car as it goes around the second curve, compared to the first?

   A. the same
   B. twice as much
   C. Four times as much
   D. ½ as much

2. A hypothetical planet has a mass of half that of Earth (which has a gravitational acceleration at the surface of *g*) and a radius of twice that of Earth. What is "*g*" on the planet?

   A. *g*
   B. *g*/2
   C. *g*/4
   D. *g*/8

3. Which of the following could be true of an object moving with constant acceleration?

   A. Moving in a circle.
   B. Increasing its velocity.
   C. Decreasing its velocity.
   D. All of the above could be true.

4. When an object experiences uniform circular motion, what is the direction of the net force on the object?

   A. outward, away from the center along the radius
   B. inward, toward the center along the radius
   C. tangential, along the path of the motion in the circle
   D. inward or outward depending on the radius of the circle

5. When a car goes around a banked curve, what is the direction of the net force on the car, and what provides that force?

   A. toward the center of the curve, only by friction
   B. toward the center of the curve, normal force from the bank and friction
   C. tangential along the curved path of the car, by friction only
   D. tangential along the curved path of the car, by a normal force from the bank

6. For a satellite in orbit, what provides the centripetal force for the orbit?

   A. gravitational attraction of the satellite on the Earth
   B. normal force
   C. gravitational attraction of the Earth on the satellite
   D. inertia of the satellite

7. Which of the following is <u>not</u> a true statement regarding a satellite in circular orbit at constant speed around a planet?

   A. The centripetal force on the satellite is always toward the center of the planet.
   B. The satellite exerts the same amount of force on the planet as the planet exerts on the satellite.
   C. The force the satellite exerts on the planet is in the same ratio to the planet's force on the satellite as the ratio of their masses.
   D. The acceleration of the satellite as it moves in its orbit is not zero.

8. An object with a mass of 2,000 kg moves with a constant speed of 20 m/s on a circular track of radius 100 m. What is the magnitude of the acceleration of the object, in m/s^2?

   A. 400
   B. 8,000
   C. 40
   D. 4

Projectile Motion

Projectile Motion is the motion of an object projected into the air near the Earth's surface under the influence of gravity. If it is projected straight upward or downward, the motion is vertical and the path will be linear, but if it is projected at an angle, the path will be a parabola. If the path is parabolic, the object in motion has two components—horizontal and vertical. The horizontal and vertical components of the motion occur simultaneously, so the time for both is the same.

> Think of projectile motion as two independent motions with separate equations in each dimension—one for the motion in the horizontal direction and one for the motion in the vertical direction. The time for each motion is the same.

Vertical component of projectile motion: In the absence of air friction, the only force exerted on an object in projectile motion is the gravitational force downward. Thus, the acceleration in the vertical direction is g, which is 9.8 m/s² downward. If the launch angle θ is measured from the surface, use $\mathbf{v} \sin \theta$ for the initial vertical velocity.

$$\mathbf{v}_y = \mathbf{v}_{oy} + \mathbf{a}t$$
$$\Delta y = \mathbf{v}_{oy}t + \tfrac{1}{2}\mathbf{a}t^2$$

Horizontal component of projectile motion: In the absence of air friction, there is no force exerted on an object in projectile motion—and no acceleration. If the launch angle θ is measured from the surface, use $\mathbf{v} \cos \theta$ for the initial horizontal velocity.

$$\mathbf{v}_x = \mathbf{v}_{ox}$$
$$\Delta x = \mathbf{v}_{ox}t$$

> If the launch of a projectile takes place horizontally (such as rolling a ball from the edge of a level table), then the initial horizontal launch velocity is just $\mathbf{v}$, and the initial vertical velocity is zero.

EXAMPLE—Graphs of Projectile Motion

The two graphs below represent the vertical displacement versus time and vertical velocity versus time for a projectile launch from the ground. (a) At what time does the projectile change direction? (b) How does the acceleration on the projectile's trip upward compare to the acceleration on the way back down? (c) What would a graph of the projectile's horizontal velocity versus time look like? (d) What would a graph of the projectile's horizontal displacement versus time look like?

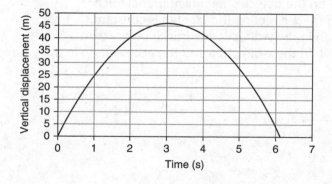

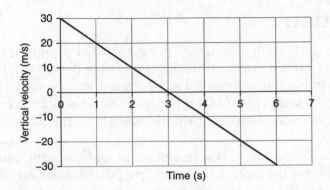

Solution:

(a) The projectile reaches its peak at $t = 3$ s, so it turns at that time to fall downward.

(b) The acceleration of the projectile is the same throughout its trip—9.8 m/s/s downward as it moves upward (which slows it down) and 9.8 m/s/s downward as it comes back down (which speed it up as it falls).

(c) There is no force and no acceleration in the horizontal direction, so the horizontal velocity is constant. The graph would just be a straight horizontal line.

(d) Since the horizontal velocity is constant, as time increases, the horizontal displacement will increase proportionally. So the graph will be a straight line with a positive slope.

EXERCISE 4.4

1. You throw a ball at an upward angle into the air, and it moves in a parabolic motion and then hits the ground. These are all true statements about that projectile motion, but which of the following is the best example of the application of Newton's first law of motion?

   A. The ball continues to move at constant speed horizontally after it leaves your hand.
   B. Your hand exerts a force on the ball as you accelerate it during your throw.
   C. The ball moves upward at a decreasing vertical speed until it reaches the top of its arc.
   D. The ball moves at an increasing vertical speed from the point at the top of its arc until it hits the ground.

2. Which statement is <u>not</u> true of the motion of an object in projectile motion? (Assume no air friction.)

   A. The only force on the object is the gravitational force.
   B. The velocity in the x-direction (horizontal) is constant.
   C. The acceleration in the y-direction (vertical) is constant.
   D. The acceleration in the x-direction and the acceleration in the y-direction are equal.

3. When any object is projected into the air from the ground at an angle so that it takes a parabolic path, the vertical (y-direction) displacement of the object for the entire trip is:

   A. Zero
   B. Always equal to the horizontal (x-direction) displacement
   C. Always less than the horizontal (x-direction) displacement
   D. Always more than the horizontal (x-direction) displacement

   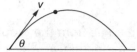

4. When any object is projected into the air from the ground at an angle so that it takes a parabolic path, at the very peak of the path (excluding an air friction):

   A. The object is moving at constant velocity horizontally (x-direction) and zero velocity vertically.
   B. The object comes to a complete stop at that instant.
   C. The object is moving at constant velocity vertically (y-direction) and zero velocity horizontally.
   D. The object is still accelerating both vertically and horizontally.

5. The path of an object's motion will be parabolic if it undergoes:

   A. constant velocity in the x-direction and in the y-direction
   B. constant acceleration in the x-direction and in the y-direction
   C. constant acceleration in one dimension and constant velocity in the second dimension
   D. increasing acceleration in one dimension

6. In a projectile motion experiment in the lab, assume a ball rolls across a level table and leaves the table at a velocity horizontally of 2.0 m/s. The table height (h) is 1.0 m. Calculate the distance (X) from the base of the table the ball will land on the floor.

   A. 0.40 m
   B. 0.63 m
   C. 0.90 m
   D. 1.2 m

7. A demonstration is set up where a metal bar holding two steel balls is attached to the top of a building. When the bar is hit sideways with a hammer, one ball flies off horizontally and one ball drops to the ground. If they leave the bar at the same moment, which ball will hit the ground first (assuming no air friction)?

   A. The ball hit horizontally will hit the ground first.
   B. The ball that just drops vertically will hit the ground first.
   C. Both balls will hit the ground at the same instant.
   D. Which ball hits the ground first depends on the masses of the steel balls.

8. A block slides across a table and off of the edge. What are the horizontal and vertical components of the block's acceleration from the time it leaves the table until it hits the floor?

   A. $a_H = 0$, $a_v = 9.8$ m/s^2
   B. $a_H = 9.8$ m/s^2, $a_v = 0$
   C. $a_H = 4.9$ m/s^2, $a_v = 4.9$ m/s^2
   D. $a_H = 4.9$ m/s^2, $a_v = 9.8$ m/s^2

9. A block slides across a flat roof that is 5 meters tall and leaves the edge moving horizontally at a speed of 2 m/s. What are the horizontal and vertical components of the block's velocity when it hits the ground below?

A. $v_H = 0$, $v_v = 2$ m/s
B. $v_H = 10$ m/s, $v_v = 10$ m/s
C. $v_H = 2$ m/s, $v_v = 10$ m/s
D. $v_H = 2$ m/s, $v_v = 14$ m/s

10. The drawing shows the path of a projectile that was launched at an angle from the ground at point A and lands on the ground again at E. Assuming negligible air friction, at which points would you find an acceleration of zero, a maximum speed, and maximum height?

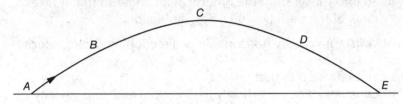

	$a = 0$	$v = $ max	height $=$ max
A.	B	E	E
B.	No point	C	D
C.	C	A	C
D.	No point	A	C

Work and Energy

- Mechanical Energy
- Work
- Work-Energy Theorem
- Conservation of Energy
- Power

Mechanical Energy

Kinetic energy is energy of motion. It is a scalar quantity, measured in joules, that describes the energy of an object with mass m moving at velocity v. Mass is measured in kilograms, and velocity is measured in m/s.

$$KE = \tfrac{1}{2}m\mathbf{v}^2$$

Potential energy is the stored energy of a system, such as the potential energy of an Earth-object system or the potential energy of a mass-spring system. For example, a person standing on the surface of the Earth has potential energy relative to the Earth due to the masses of both the person and the Earth and the distance between the center of Earth and the center of the person (R). This type of potential energy is gravitational potential energy:

$$PE_G = \frac{Gm_1m_2}{R}$$

If the person climbs a set of stairs and increases the distance R, they have increased the gravitational potential energy of the system of two objects. This increase in potential energy can be determined by calculating a new PE and then subtracting to find the difference.

Other forms of potential energy, which will be reviewed later, include spring potential energy of a mass-spring system, electric potential energy due to two charged particles, and electric potential energy of a charged capacitor:

$$PE_s = \tfrac{1}{2}kx^2 \qquad PE_E = \frac{kq_1q_2}{R} \qquad PE_C = \tfrac{1}{2}CV^2$$

NGSS HS-PS3-1

When using these formulas, energy is in joules when mass is in kilograms, height and radius are in meters, and g is in m/s^2.

A single object cannot by itself possess potential energy. The potential energy changes as relative positions of objects within the system change. For example, a satellite moving around the Earth has gravitational potential energy relative to Earth—and that potential energy changes as the satellite's distance from Earth changes.

$$U_G = PE_G = mgh$$

For problems involving small displacements near the Earth's surface, we can define changes in gravitational potential energy, using any reference line we choose as the "zero" point:

$$\Delta U = mg\Delta h$$

The potential energy changes of a mass on a pendulum are really gravitational potential energy changes of the pendulum "bob" with respect to the Earth. We can define the lowest point of the pendulum bob's swing as the *zero point* and the highest point to which the pendulum swings as the point where the bob has its highest potential energy.

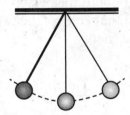

Elastic Potential Energy: The potential energy of a mass attached to a spring depends on the **spring constant**, k, and the amount the spring is stretched or compressed from its equilibrium position. The change in potential energy of a spring varies with the displacement from equilibrium, since the spring force is not a constant force (i.e., it takes more energy to stretch a spring the same amount as it becomes more stretched).

$$PE = U_{spring} = \tfrac{1}{2}k(\Delta x)^2$$

Remember: The spring constant is the proportionality constant for the direct relationship between the force of the spring and the amount the spring is stretched or compressed, with the force in the opposite direction of the spring's displacement:

$$F = -kx$$

EXAMPLE—Conservation of Energy on a Ramp

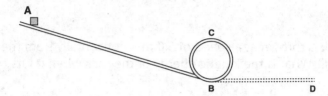

A small block is released from rest on a steel track at point **A**, which is 0.8 meter higher than point **B**. The block slides through the loop and then leaves the track at point **D** (shown above). The radius of the loop is 0.2 meter. (a) Calculate the speed of the block at point **B**. (Ignore friction.) (b) Calculate the speed of the block at point **C**. (Ignore friction.)

Solution:

(a) Without friction, mechanical energy is conserved, so potential energy at point A is converted to kinetic energy at point B with no "loss" in mechanical energy:

$$|\Delta U| = |\Delta K|$$

$$\cancel{m}gh = \tfrac{1}{2}\cancel{m}v^2$$

$$v_B = \sqrt{2gh} = \sqrt{(2)(9.8 \text{ m/s}^2)(0.8 \text{ m})} = 4.0 \text{ m/s}$$

(b) Since the radius of the loop is 0.2 m, point **C** is at a height of 0.4 m relative to point **B.** Part of the original potential energy at point A is now in the form of kinetic energy at point C and part is potential energy. There is still no "loss" to work against friction force.

$$U_A = U_C + K_C$$

$$\cancel{m}gh_A = \cancel{m}gh_C + \tfrac{1}{2}\cancel{m}v_C^2$$

$$(9.8 \text{ m/s}^2)(0.8 \text{ m}) = (9.8 \text{ m/s}^2)(0.4 \text{ m}) + \tfrac{1}{2}\cancel{m}v_C^2$$

$$v_C = 2.8 \text{ m/s}$$

1. A pendulum is pulled back and released from a position 0.2 m higher than its position at the bottom of its swing. What is the speed of the pendulum bob as it moves through the equilibrium position at the bottom of its swing after it is released?

   A. 0.5 m/s
   B. 1.0 m/s
   C. 1.5 m/s
   D. 2.0 m/s

2. A 0.5-kg rock is thrown at a speed of 5.0 m/s horizontally from the top of a building 20 meters tall. What is the kinetic energy of the rock when it hits the ground?

   A. 6.3 J
   B. 104 J
   C. 112 J
   D. 160 J

3. An automobile with a mass of 2,000 kg accelerates from a speed of 10 m/s to a speed of 20 m/s in a time period of 6 s. What is the change in kinetic energy of the vehicle during this time period?

   A. 1,000 J
   B. 6,000 J
   C. 20,000 J
   D. 300,000 J

4. Which of the following graph shapes could represent the plot of kinetic energy versus time for an object thrown horizontally from the top of a building?

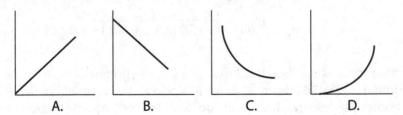

   A.   B.   C.   D.

5. A block with a mass of 2.0 kg is moving at a speed of 3.0 m/s when it slides horizontally off of a roof that is 4.0 m above the ground. The kinetic energy of the block just before it hits the ground is closest to:

   A. 9 J
   B. 20 J
   C. 78 J
   D. 87 J

6. A rock with mass m is thrown horizontally with a speed $\mathbf{v}$ from the top of a building of height h. What is the kinetic energy of the rock just before it hits the ground?

   A. mgh
   B. $\frac{1}{2} m\mathbf{v}^2$
   C. $mgh - \frac{1}{2} m\mathbf{v}^2$
   D. $\frac{1}{2} m\mathbf{v}^2 + mgh$

7. The gravitational potential energy of a system of two particles is equal to U. In a second system, the mass of each particle is twice is much and the distance between their centers is half as great. What is the gravitational potential energy of the second system in terms of U?

A. ½ U
B. 2 U
C. 4 U
D. 8 U

8. A spring with spring constant $k = 200$ N/m has an object with a mass of 40 g attached to it. Compare the potential energy of the spring/mass system when the spring is extended 0.01 m beyond the equilibrium position to the potential energy when the spring is extended twice as far, to 0.02 m beyond equilibrium.

A. 1:1
B. 1:2
C. 1:4
D. 1:16

9. A flexible plastic ruler held against the top edge of a desk is flexed sideways a distance of 5 cm, and a 0.1-kg rock is placed against the end of the ruler. When the ruler is released and the rock is propelled away from the edge of the desk, it is determined that the rock has a horizontal speed of 2.0 m/s. Determine the elastic constant, k, of the ruler.

A. 160 N/m
B. 80 N/m
C. 16 N/m
D. 8 N/m

Work

Work: An external force, $\vec{F}$, acting on an object or system through a displacement, $\vec{s}$, does **work**, W. Work is a scalar quantity measured in joules, and the calculation of work uses only the component of force acting along the line of the displacement:

$$W = Fs \cos \theta$$

The $\cos \theta$ attached to this formula applies to the angle between the direction of the force and direction of displacement. It essentially takes the component of force in the same direction as the displacement. When the displacement and force components are in the *same direction*, work is *positive*, and when the displacement and force components are in the *opposite direction*, work is *negative*. When the displacement and force components are perpendicular to each other, work is zero!

Positive work: Work done *on* a system to increase the **mechanical energy** (potential and kinetic energy) of the system. **Negative work** is done *by* the system and decreases the system's mechanical energy.

1. In which case will the most work be done by a force F in pushing a box a distance d across a level surface at constant speed?

   A. F is pushing horizontally.
   B. F is pushing downward at angle 30° to the floor.
   C. F is pushing downward at angle 60° to the floor.
   D. F is pushing vertically downward.

2. In which case is the net work positive?

   A. Work done by the tension in a string whirling a ball in a horizontal circle.
   B. Work done in lifting and lowering a set of 25-kg barbells from the floor 10 times.
   C. Work done in holding the 25-kg barbells at a constant height for 3 minutes.
   D. Work done in kicking the set of 25-kg barbells so that they begin to roll across the floor.

3. How much work (in joules) is done by the centripetal force (in this case, the friction between the tires and the road) on a 1,000-kg car moving at a constant speed of 10 m/s on a level circular track of radius 50 m?

   A. 50,000 J
   B. 20,000 J
   C. 2,500 J
   D. No work is done by the centripetal force.

4. Which of the following expressions correctly describes the work done <u>by friction</u> on a box of mass m, sliding across a level floor with coefficient of friction μ, for a distance d?

   A. $\mu m d$
   B. $-\mu m g d$
   C. $\mu m g d$
   D. $-\mu g d$

5. In which case will the most work be done by a force on a box of mass m in pushing the box a distance d across a level surface at constant speed?

   A. The force F is horizontal and parallel to the floor.
   B. The force F is at an angle 30° to the floor.
   C. The force F is at an angle 45° to the floor.
   D. The force F is at an angle 60° to the floor.

6. A spring with spring constant 200 N/m is placed on a horizontal surface and attached at one end to a wall. A block with a mass of 0.5 kg is used to compress the spring a distance of 5 cm. If the mass of the spring is negligible and the coefficient of friction between the block and the surface is 0.02, determine the speed of the block at the moment it loses contact with the spring.

   A. 0.99 m/s
   B. 0.5 m/s
   C. 0.1 m/s
   D. 0.02 m/s

7. Students want to test a small pop-up toy to determine the spring constant of its spring. They assume that the mass of the small plastic top of the toy is negligible. When the students put a 20-g glob of putty on the toy, then push the toy down 1.0 cm and release it, the spring causes the toy to move a distance upward of 8 cm. Using a preliminary calculation from this trial, what do the students predict for the spring constant?

A. 80 N/m
B. 160 N/m
C. 310 N/m
D. 480 N/m

8. What is the work done in placing a satellite of mass *m* in orbit at a distance 2*R* from the surface of Earth?

A. The work is equal to the weight of the satellite plus the kinetic energy of the satellite.
B. The work is equal to the gravitational potential energy of the satellite at that altitude above Earth's surface.
C. The work is equal to the gravitational potential energy of the satellite plus the centripetal force necessary to keep the satellite in orbit at that altitude.
D. The work is equal to the change in gravitational potential energy of the satellite from the surface to altitude plus the change in kinetic energy of the satellite.

9. In which case has a person done the most work on the system described?

A. Moving a 20-kg box at constant speed across a surface with a coefficient of friction of 0.05, using a force of 10 N for a distance of 10 m.
B. Lifting a 20-kg box onto a shelf that is 2 m high.
C. Holding a 20-kg box 2 meters above the floor for 10 minutes.
D. Pushing a 20-kg box across a frictionless surface to accelerate it from rest to 5 m/s.

10. An 11-lb bowling ball (mass about 5.0 kg) is dropped from a height of 1.0 m onto a floor. If the dent the ball makes in the floor is about 1.0 cm deep, estimate the average force the ball exerted on the floor.

A. 50 N
B. 250 N
C. 2,500 N
D. 5,000 N

Work-Energy Theorem

Work-Energy Theorem: Work done on a system by an external force changes the energy of the system: $W = \Delta K + \Delta U$

EXAMPLE (Advanced)—Work on a Box on a Ramp

An average force of 34.5 N is applied to a 3.0-kg box in the direction shown to push the box up a ramp that is inclined at an angle of 30°. (a) Calculate the change in gravitational potential energy of the box from the bottom of the ramp to the top of the ramp. (b) If the coefficient of kinetic friction between the box and ramp is 0.2, calculate the net work done by external forces to move the box to a position at the top of the ramp.

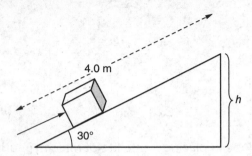

Solution:

The system under examination here is the box-Earth system, or the box in Earth's gravitational field—so anything the gravitational force does to the box does not change the total energy. The pushing force and friction force are external to this system, so they will cause changes in total energy.

(a) The pushing force will increase the gravitational potential energy, but the gravitational force is a conservative force, so the actual path doesn't matter. Simply use mgh, with h the increased distance between Earth and the box.

$$\text{Using trigonometry: } \sin 30° = \frac{h}{4} \qquad h = 4 \sin 30°$$

$$\Delta U_G = mgh = (3.0 \text{ kg})(9.8 \text{ m/s}^2)(4.0 \sin 30°) = 59 \text{ J}$$

(b) The friction force is nonconservative, so the work done by the friction force is equal to the friction force times the entire distance the friction force is exerted—or the distance the box is moved in contact with the ramp. The pushing force also acts through the entire distance, l. Start with a free-body diagram of the box on the ramp, and find the components of the gravitational force in order to determine the normal force.

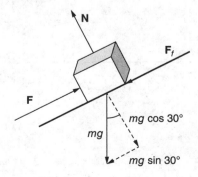

Substitute and calculate the net work:

$$\mathbf{F}_f = \mu N \qquad \text{and} \qquad N = mg \cos 30°$$

so $\Sigma W = (F_{net})(l) = (34.5 - F_f - mg \sin 30°)(l) = (34.5 - \mu\, mg \cos 30° - mg \sin 30°)(l)$

$\Sigma W = [34.5 - (0.2)(3.0\ \text{kg})(9.8\ \text{m/s}^2)(\cos 30°) - mg \sin 30°)](4.0\ \text{m}) = 59\ \text{J}$

This confirms the **work-energy theorem**—that the net work done on an object is equal to the change in energy.

Note that work done by the friction force is negative because the friction-force vector is in the opposite direction of the displacement up the ramp. Additionally, friction causes mechanical energy of the system to be "lost"—or dissipated to thermal energy.

EXAMPLE—Work on a Spring

A spring with a spring constant value of 75 N/m is pulled back a distance $x = 10$ cm and held in that position. (a) How much work is done on the spring by the external force? (b) What is the potential energy of the spring when held in that position? (c) What would be the average force required to pull the spring back a distance of 20 cm?

Solution:

(a) From the work-energy theorem, the work is equal to the potential energy of the spring in the pulled-back position.

$$W = \Delta U = \tfrac{1}{2}kx^2$$
$$W = \tfrac{1}{2}(75\ \text{N/m})(0.10\ \text{m})^2 = 0.38\ \text{J}$$

The work equation "force times distance" can't be used with a spring to find work or change in energy, because the spring force varies with the distance the spring is stretched. It is not a constant force.

(b) The potential energy is equal to the work done by the outside force in stretching the spring: 0.38 J

(c) Using the Hooke's law equation:

$$\vec{F}_{applied} = k\Delta\vec{x} = (75\ \text{N/m})(0.20\ \text{m}) = 15\ \text{N}$$

Don't forget that the spring constant or elastic constant, k, has units newtons per meter that indicate the resistance of the spring to being stretched or compressed. This is the proportionality constant between force and displacement of the spring.

EXAMPLE—Work-Energy Theorem with Friction

A brick with a mass of 1.5 kg is sliding at a speed of 2.0 m/s across a frictionless floor when it slides up onto a ramp, where the coefficient of kinetic friction between the brick and ramp is 0.8. The ramp has a length of 2.2 m and is inclined at an angle of 30°. How far up the length of the ramp does the brick slide before it comes to a stop?

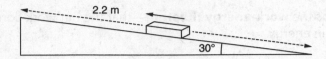

Solution:

Using the work-energy theorem, we see that at the point the brick comes to a stop, its kinetic energy has been converted to gravitational potential energy and thermal energy due to work done against friction. Items to consider first:

1. The gravitational force is a conservative force, so we use the height of the ramp, h, to calculate potential energy.

2. The friction force is nonconservative, so we use the <u>entire</u> distance moved, d, to calculate friction force.

3. Use the formula $F_f = \mu N$ to calculate friction force.

4. From the free-body diagram of the brick on the ramp, we see that the normal force is equal to mg cos 30°.

5. Work done against friction is $F_f \times d$.

6. We don't make the value for g negative, because we're calculating energy, which is scalar.

Substituting these into the equation, we get:

$$|\Delta K| = |\Delta U| + E$$

$$\tfrac{1}{2}m(\Delta v)^2 = mg\Delta h + (\mathbf{F}_f)(d)$$

$$\tfrac{1}{2}\cancel{m}(v_f - v_o)^2 = \cancel{m}(9.8 \text{ m/s}^2)(\Delta h) + (\mu \cancel{m}g)(\cos\theta)(d)$$

$$\tfrac{1}{2}(0 - 2.0 \text{ m/s})^2 = (9.8 \text{ m/s}^2)(d \sin 30°) + (0.8)(9.8 \text{ m/s}^2)(\cos 30°)(\mathbf{d})$$

$$2 = 4.9\, d + 6.79\, d$$

$$d = 0.17 \text{ m} \quad \text{or} \quad 17 \text{ cm}$$

EXAMPLE

Analyze the following data from an experiment with a spring to determine: (a) the spring constant (k); (b) the work done by an external force in stretching the spring 2.0 cm from its equilibrium position; and (c) the potential energy of the spring when it is stretched 2.0 cm from equilibrium.

Spring Displacement, Δx (cm)	Applied Force, F (N)
1.2	58.0
2.0	99.0
2.5	125
3.2	161

Solution:

Rather than using data to make a calculation, graph the data to analyze trends. Since $F_{applied} = kx$, the slope of this linear plot will be the spring constant.

Area under the plot line with be equal to the work done on the spring to stretch it, which is equal to the spring's potential energy in each case. The area is a triangle: $A = \frac{1}{2} bh$. In this case, $A = \frac{1}{2} Fx$. Since $F = kx$, this formula becomes:

Area = $\frac{1}{2} \times$ base $\times$ height = $\frac{1}{2} F*x = \frac{1}{2} (kx)x = \frac{1}{2} kx^2$, which is the formula for potential energy of a spring!

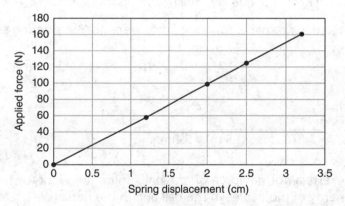

(a) Slope $= k = \dfrac{\Delta y}{\Delta x} = \dfrac{150-100}{3-2} = 50$ N/cm or 5,000 N/m

(b) Work $=$ area $= \frac{1}{2}(0.02$ m$)(100$ N$) = 1$ J

(c) Potential energy $= W = 1$ J

EXAMPLE—Energy Conversion with a Falling Ball

You lift a 4-kg bowling ball from the floor to a height of 1.5 m. (a) How much work have you done on the ball? (Note whether it is positive or negative.) (b) How much potential energy does the ball now have, relative to the floor? (c) Calculate the speed of the bowling ball at the instant just before it hits the floor. (e) After the ball hits the floor and stops, explain what has happened to the energy it had before it hit.

Solution:

(a) You are exerting an external force on the ball, so you change its potential energy with respect to Earth. By lifting the ball and increasing the distance between the ball and Earth, you increase the potential energy of that system.

$$W = \Delta U = mgh = (4 \text{ kg})(9.8 \text{ m/s}^2)(1.5 \text{ m}) = 59 \text{ J}$$

(b) The increase in potential energy with respect to its previous position is equal to the work or the increase in potential energy, which is 59 J.

(c) Mechanical energy remains constant, since we're not considering air drag, so:

$$|\Delta U| = |\Delta K|$$

$$mgh = \tfrac{1}{2}m\mathbf{v}^2$$

$$\mathbf{v} = \sqrt{2gh} = \sqrt{(2)(9.8 \text{ m/s}^2)(1.5 \text{ m})} = 5.4 \text{ m/s}$$

(d) The floor exerts a normal force on the ball, doing the work to stop it. This work becomes thermal energy of molecules in the ball and in the floor. (We'll learn in a later chapter that this increase in thermal energy means the molecules within the ball and floor now have greater kinetic energy.)

1. If a ball is dropped, hits the floor, and bounces back upward:

   A. It has positive kinetic energy just before it hits the floor and negative kinetic energy after it bounces upward from the floor.
   B. The floor does negative work on the ball to slow it down as it first hits.
   C. The floor does negative work on the ball to accelerate it back upward after it hits.
   D. The gravitational force does negative work on the ball to accelerate it when it is dropped.

2. Which of the following is an accurate statement regarding nonconservative forces?

   A. The gravitational force is nonconservative because it increases on an object as the object moves closer to the center of Earth.
   B. The gravitational force is nonconservative because the path taken by an object is always in the same direction as the gravitational force exerted on the object.
   C. The friction force is nonconservative because mechanical energy is converted to thermal energy.
   D. The friction force is nonconservative because work done by friction is always independent of path taken.

3. A ball is placed on a ramp and allowed to roll to the bottom of the ramp and onto a level table, where the velocity of the ball is determined by how far the ball lands from the edge of the table on the floor below. The kinetic energy calculated at the bottom of the ramp is significantly less than the gravitational potential energy change from the top to the bottom of the ramp. What is the best reason for this discrepancy?

   A. Work done by the gravitational force on the ball
   B. Work done by the normal force on the ball
   C. Work done by the friction force on the ball
   D. Transfer of thermal energy to the ball from the ramp

4. A 2-kg object slides from rest a distance of 30 m down a snow-covered hill (i.e., no friction) to a point that is 10 m lower on the hill, where it stops. What is the magnitude of work done by the gravitational force on the object?

   A. 20 J
   B. 60 J
   C. 200 J
   D. 600 J

5. A satellite is in a nearly circular orbit around the Earth. What provides the work necessary to maintain the kinetic energy of the satellite?

   A. The gravitational force of the Earth on the satellite
   B. The centripetal force on the satellite
   C. The gravitational force of the satellite on the Earth
   D. No work is required.

Conservation of Energy

Energy is conserved within a defined system if no energy enters or leaves the system and/or if no work is done **on or by the system.**

> **Conservative Force:** Force such as the gravitational force that does not cause loss of mechanical energy to thermal energy.
>
> **Thermal Energy:** Energy due to motion of molecules. Frictional "losses" are just transfer of mechanical energy of objects within a system to increased motion of molecules within those objects.
>
> **Non-Conservative Force:** Force, such as friction force, that causes "loss" of mechanical energy to thermal energy.

It's important to define the **system** under examination. Energy transfers within a system, changes in configuration within a system, or forces acting within a system do not change the total energy of the system.

EXAMPLE (Advanced)—Energy of a Pendulum

Suppose a bowling ball pendulum has a mass of 4.0 kg (neglecting the mass of the chain) and the length of the chain from its pivot to the center of the bowling ball is 2.0 m. When the bowling ball is pulled back an angle of 20° and released, what will be its speed when it first comes back to the bottom of its swing?

Solution:

If friction is negligible, the gravitational potential energy of the ball just before its release is equal to the kinetic energy of the ball at the bottom of its swing, because mechanical energy is conserved. The only difficulty here is in calculating the potential energy of the ball at the point where it is a distance h above its lowest point.

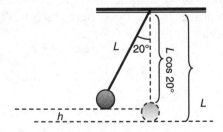

When the ball is lifted before release, it is a height h above its lowest point. Using some simple trigonometry: $h = L - L \cos 20°$,

$$|\Delta U| = |\Delta K|$$
$$mgh = \tfrac{1}{2} m v^2$$
$$g(L - L \cos 20°) = \tfrac{1}{2} v^2$$
$$(9.8 \text{ m/s}^2)(2 - 2 \cos 20°) = \tfrac{1}{2} v^2$$
$$v = \sqrt{(2)(9.8 \text{ m/s}^2)(2 - 2 \cos 20°)} = 1.5 \text{ m/s}$$

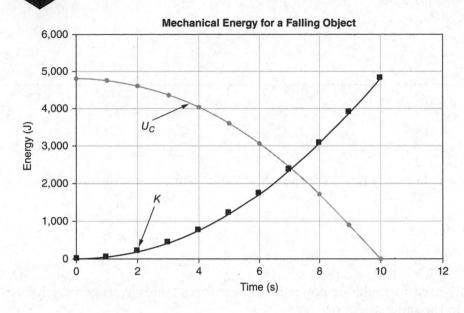

Mechanical Energy for a Falling Object

1. A brick is dropped from a height of 490 m. Use the graph to determine the mass of the brick from the information about its energy. (Neglect the effects of air friction and assume *g* does not change appreciably.)

   A. 1 kg
   B. 2 kg
   C. 3 kg
   D. 4 kg

2. Determine the kinetic energy and total energy of the brick 7 s after it is dropped.

   A. $KE = 2{,}450$ J, $E = 4{,}900$ J
   B. $KE = 4{,}900$ J, $E = 4{,}900$ J
   C. $KE = 4{,}900$ J, $E = 2{,}450$ J
   D. $KE = 0$, $E = 4{,}900$ J

3. Use the information on the graph to compare the kinetic energy of the falling object after 2 s of fall to kinetic energy after 4 s of fall.

   A. about 1:1
   B. about 1:2
   C. about 2:1
   D. about 1:4

4. A 2.0-kg ball dropped from a height of 4.0 m loses 10% of its mechanical energy to thermal energy when it hits the floor. Determine the kinetic energy of the ball just after it hits the floor.

   A. 78.4 J
   B. 70.6 J
   C. 7.84 J
   D. 3.84 J

5. A 2.0-kg ball dropped from a height of 4.0 m loses 10% of its mechanical energy to thermal energy when it hits the floor. Determine the height to which the ball will bounce after it hits the floor the first time.

A. 4.0 m
B. 3.6 m
C. 2.2 m
D. 2.0 m

Power

Power: The rate at which energy is used or the rate at which work is done, measured in joules per second or watts.

$$P = \frac{E}{t} = \frac{W}{t}$$

Another commonly used formula for power is **average force times average velocity**, with the force and velocity in the same direction:

$$P = Fv$$

If you examine the units on this formula, they are newtons times meters per second, which is newton-meters per second—the same units as on energy divided by time or watts.

Energy: Equal to average power times time. If power is in watts and time is in seconds, energy is in joules. However, for commercial energy use, power is measured in kilowatts and time is in hours, so energy is in kilowatt-hour (kWh). Commercial customers are then billed per kilowatt-hour of energy usage, depending on region. Fifteen cents per kilowatt-hour is a common value.

EXERCISE 5.5

1. In which situation is the most average power required?

A. Lifting a 5-kg block to a height of 2 m in 2 s
B. Pushing a block across a level surface with a net force of 10 N at a velocity of 3 m/s
C. Changing the kinetic energy of a rolling wheel from 15 J to 55 J in 20 s
D. Burning a 10 W light bulb for 20 h

2. Which statement correctly relates work and power?

A. Work may be either positive or negative, but power must be positive.
B. Work and power must both always be positive.
C. If work is positive, power must be positive.
D. If work is negative, power must still be positive.

3. A 2200-kg vehicle is accelerated from rest to a speed of 30 m/s in 40 s along a level road. What is the net work done on the vehicle?

    A. 990,000 J
    B. 49,500 J
    C. 24,740 J
    D. 825 J

4. A 2200-kg vehicle is accelerated from rest to a speed of 30 m/s in 40 s along a level road. What is the average power generated to accelerate the vehicle during that 40 s?

    A. 990,000 W
    B. 49,500 W
    C. 24,740 W
    D. 825 W

5. An average force of 0.2 N is exerted on a 2-kg object to accelerate it from a speed of 2 m/s to 3 m/s in a time interval of 10 s. Determine the average power.

    A. 0.5 W
    B. 2.0 W
    C. 5.0 W
    D. 20 W

Momentum

- ■ Linear Momentum
- ■ Force and Impulse
- ■ Conservation of Linear Momentum

Linear Momentum

Conservation of linear momentum is another conservation law that applies in many situations. This conservation law, along with conservation of kinetic energy, helps us to understand how objects and systems interact with each other—particularly during collisions. This unit helps to clarify, through examples, when both laws can be applied and when only momentum is conserved. Newton's second law of motion, in terms of force and impulse, will be discussed as we examine momentum changes.

Linear momentum, p: A vector quantity that describes the motion of an object; it is the product of the object's mass, *m*, and linear velocity, **v**. *The momentum is in the same direction as the velocity.*

$$\mathbf{p} = m\mathbf{v}$$

Momentum in one dimension: Calculations can just use positive and negative values to indicate the vector directions.

Momentum in two dimensions: The calculations are separate for each dimension, *x* and *y*, using vector components. This is usually beyond an introductory course.

Center of Gravity or Center of Mass: The "balance point" in a gravitational field or the point in a system where the total mass of the system is concentrated. This is particularly important because forces that are applied through the center of mass will produce linear motion instead of rotational motion.

EXAMPLE—Calculate Center of Mass of a System

Four objects are placed on a sheet of paper. We want to find the center of mass of the four objects. First, we draw axes to find locations of the objects. They are as follows:

Object A 2 g located at 3, 7

Object B 6 g located at –2, 6

NGSS HS-PS2-2 and HS PS 3-3 and HS PS 3-2

Object C 5 g located at 4, −8

Object D 10 g located at −2, −6

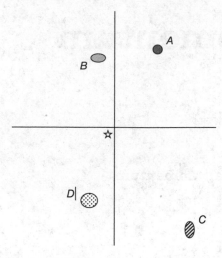

Calculate the *x*-location and the *y*-location of the center of mass:

$$x_{com} = \frac{m_1 x_1 + m_2 x_2 + m_3 x_3 + m_4 x_4}{\Sigma m} = \frac{(2g)(3) + (6g)(-2) + (5g)(4) + (10g)(-2)}{23g} = -0.26$$

$$y_{com} = \frac{m_1 y_1 + m_2 y_2 + m_3 y_3 + m_4 y_4}{\Sigma m} = \frac{(2g)(7) + (6g)(6) + (5g)(-8) + (10g)(-6)}{23g} = -2.17$$

The small star on the sheet shows the location of the center of mass of the system of coins.

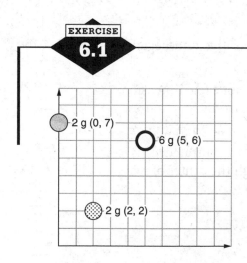

EXERCISE
6.1

1. The diagram above shows the positions of three coins on an *x-y* grid. Use their masses and positions to determine the coordinates of the center of mass of the system of coins.

   A. 2.2, 4.0
   B. 3.4, 4.0
   C. 2.8, 5.0
   D. 3.4, 5.4

2. An object has a mass of quantity *m* and a momentum of quantity *p*. If the mass and the velocity of the object double, what happens to momentum and kinetic energy?

   A. Momentum doubles and kinetic energy doubles.
   B. Momentum doubles and kinetic energy is eight times as much.
   C. Momentum is four times as much and kinetic energy doubles.
   D. Momentum and kinetic energy are both four times as much.

3. What is the momentum of a 2,000-kg vehicle moving at 60 mi/h?

   A. 53,600 J/s
   B. 53,600 kg-m/s
   C. 53,600 N-s
   D. 53,600 kg/m

4. Which of the following is true regarding the vector nature of momentum and kinetic energy?

   A. Both are vectors.
   B. Both are vectors.
   C. Momentum is a vector but kinetic energy is a scalar.
   D. Momentum is a scalar but kinetic energy is a vector.

5. Which of the following has the largest magnitude of momentum?

   A. A 5,000-kg railroad car moving at 10 miles/hour
   B. A 2,000-kg car moving at 70 miles/hour
   C. An electron moving at 2×10^6 m/s
   D. A 2,000-kg car moving at 20 m/s

Force and Impulse

Impulse, **J** or **Δp**, is a vector quantity that is equal to change in momentum:

$$\mathbf{J} = \Delta\mathbf{p} = m\Delta v$$

The linear momentum of an object or system of objects does not change unless an *external force* is applied to the object or system. The amount of force required to change momentum or the amount of force exerted by an object as it changes momentum depends upon the rate of momentum change:

$$\mathbf{F} = \frac{\Delta\mathbf{p}}{\Delta t}$$

Likewise, an impact in which momentum is changed will exert a force. Examining this equation, an impact or change in momentum over a short period of time will exert a larger force. *The direction of the force exerted on an object or system is the same as the direction of change in momentum or direction of impulse on the object.*

In the diagram below, the cart will roll down the ramp and along a track with negligible friction to collide with a stationary force sensor.

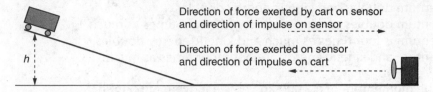

The force exerted by the sensor to the left on the cart changes the cart's **momentum** toward the left in the diagram. This is an impulse to the left.

Newton's second law of motion originally defined force in terms of rate of **change in momentum**— much like the equation above instead of $\sum F = ma$, with which we are much **more familiar**. We can show that if $F = ma$ and $a = \Delta v/\Delta t$, then $F = m(\Delta v/\Delta t)$, which is the same as $F = \Delta p/\Delta t$.

Newton's third law of motion reminds us that if there is a force exerted on **an object or** system, an equal amount of force exerted by the system in the opposite direction.

EXAMPLE—Force versus Time Graph

Use the following Force versus Time graph to determine the speed of a **football after** a kick. The mass of the football is 420 g, and we assume it starts from rest.

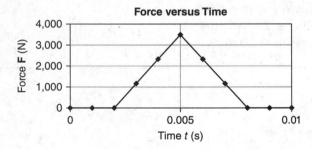

Solution:

The area under the force versus time graph line is equal to the change in **momentum** of the object being examined. The area is a triangle, so area is "1/2 base times **height." This is** really the average force multiplied by time of interaction.

$$\bar{F} = \frac{\Delta p}{\Delta t} = \frac{m\Delta v}{\Delta t}$$
$$m(v_f - v_i) = F\Delta t$$
$$(0.42\ kg)(v_f - 0) = \text{area} = \tfrac{1}{2}(0.006\ s)(3500\ N)$$
$$v_f = 25\ \text{m/s}$$

The football leaves the kicker's foot at 25 m/s after the kicker exerts an average **force of 1,750 N** on the ball over a time of 6 ms.

Don't forget: Momentum is a vector that has direction, so a positive **momentum** is in one direction, and negative momentum is in the opposite direction.

EXAMPLE—Impulse from a Graph

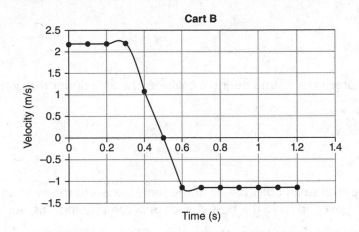

Cart B

The graph above is a computer screenshot created by a motion sensor plotting the velocity of Cart B as it collides with another cart. The mass of Cart B is 1.5 kg. Determine: (a) the approximate time during which the two carts are in contact; (b) the change in velocity of Cart B; (c) the force exerted on Cart B.

Solution:

(a) From the graph, the velocity starts changing at 0.3 s and levels off again at 0.6 s, so the time for the collision is 0.3 s.

(b) The change in velocity is from about 2.2 m/s to about −1.2 m/s.

$$\Delta v = v_f - v_o = (-1.2 \text{ m/s}) - (+2.2 \text{ m/s}) = -3.4 \text{ m/s}$$

(c) $$\mathbf{F} = \frac{\Delta \mathbf{p}}{\Delta t} = \frac{m\Delta v}{t} = \frac{(1.5 \text{ kg})(-3.4 \text{ m/s})}{0.3 \text{ s}} = -17 \text{ N}$$

The change in velocity of the cart is negative, so the force that was exerted on it to make that change is negative. Of course, Cart B exerted an equal and opposite force on the cart that it hit.

EXERCISE

6.2

1. A cart rolls across a level floor and strikes a wall elastically at 10 m/s. On a second trial, the cart strikes the wall elastically at 20 m/s. Assuming that the contact time between the cart and the wall is the same in both cases, compare the force the cart exerts on the wall in the second trial to the force in the first trial.

   A. the same
   B. ½ as much
   C. twice as much
   D. four times as much

2. The rate at which momentum changes is:

   A. impulse
   B. kinetic energy
   C. force
   D. acceleration

3. On a graph of Force versus Time during a collision, the area under the graph line is:

   A. work
   B. kinetic energy
   C. impulse
   D. momentum

4. You are given several small objects—all of the same mass—to throw at an upright block of wood to knock it over. To have the best chance of accomplishing this, you will choose:

   A. a small dart thrown at the top of the block so that it sticks into the wood
   B. a small dart thrown at the bottom of the block so that it does not stick
   C. a small, elastic rubber ball thrown at the base of the block so that it bounces
   D a small, elastic rubber ball thrown at the top of the block so that it bounces

5. The plot below shows the force exerted on a tennis ball during a collision with a tennis racket. The area between the plot line and the time axis represents:

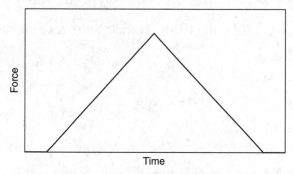

   A. The acceleration of the tennis ball while it's in contact with the racket
   B. The change in momentum of the tennis ball when it hits the racket
   C. The change in velocity of the tennis ball when it hits the racket
   D. The work done by the tennis ball on the racket when it hits the racket

6. A car moving at a speed v is brought to rest in time t by a certain force F_1. To stop the car in one-half the time—or in a time ½ t—what force F_2 is required?

   A. Force F_2 is twice as large as F_1.
   B. Force F_2 is half as large as F_1.
   C. Force F_2 is one-fourth as large as F_1.
   D. The force required to stop the car is the same in both situations.

7. When automobiles have to stop quickly, the injury to occupants can be reduced by using airbags. What is the primary mechanism by which airbags reduce injury?

   A. The airbag increases the time it takes to stop the person, so the force on the person is less.
   B. The airbag increases the distance over which the person comes to a stop, so the acceleration is smaller.
   C. The airbag is soft, so it absorbs the force of the impact.
   D. The airbag applies a force on the person in the opposite direction that the car is exerting its force on the person, so the net force on the person is less.

8. A baseball is thrown toward a target at a speed *v* and exerts a force *F* on the target when it collides with it. What is the force exerted on the target if the speed of the baseball is doubled on the second throw?

  A. ½ F
  B. F
  C. 2 F
  D. 4 F

Conservation of Linear Momentum

During an interaction among objects in which there is no external applied force, the total momentum of the system does not change. This is the **law of conservation of linear momentum.** Since momentum is a vector, components of the motion in each direction must be considered separately:

$$\Sigma p_{x(\text{before})} = \Sigma p_{x(\text{after})}$$

$$\Sigma p_{y(\text{before})} = \Sigma p_{y(\text{after})}$$

$$\Sigma p_{z(\text{before})} = \Sigma p_{z(\text{after})}$$

Elastic Collision: Interaction between objects (in the absence of an external force) during which both linear momentum and kinetic energy are conserved.

> Special case: It's not difficult to prove that moving objects of equal mass that collide elastically will exchange velocities. A moving object that collides head-on and elastically with a stationary object of the same mass will stop—and send the stationary object on with the same velocity. Additionally, if a moving object collides off-center and elastically with a stationary object of the same mass, they will both move away from—and at a right angle to—each other.

Inelastic Collision: Interaction between objects (in the absence of an external force) during which linear momentum is conserved but kinetic energy is not conserved. In a **totally inelastic** collision, the objects stick together during collision and behave as one object afterward.

EXAMPLE (Advanced)—Elastic Collision in Two Dimensions

An object with mass m_1 moving with speed *v* collides elastically with an object that has mass m_2 and is initially at rest. Write equations that could be used to solve for the velocities of both objects after the collision.

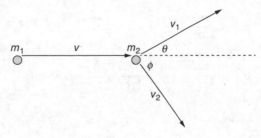

Solution:

Momentum is a vector quantity, so component equations must be used in each dimension.

Step 1. Write the equation for conservation of linear momentum in the *x*-direction:

$$\Sigma p_{x(o)} = \Sigma p_{x(f)}$$

$$m_1 v + 0 = m_1 v_1 \cos \theta + m_2 v_2 \cos \phi$$

Step 2. Write the equation for conservation of linear momentum in the *y*-direction:

$$\Sigma p_{x(o)} = \Sigma p_{x(f)}$$

$$m_1 v + 0 = m_1 v_1 \sin \theta + m_2 v_2 \sin \phi$$

Step 3. Write the equation for conservation of kinetic energy.

Since kinetic energy is not a vector, the kinetic energy equation does not use components. Total energy before collision must equal total energy after collision if the collision is elastic.

$$\Sigma KE_o = \Sigma KE_f$$

$$\tfrac{1}{2} m_1 v^2 + 0 = \tfrac{1}{2} m_1 v_1^2 + \tfrac{1}{2} m_2 v_2^2$$

EXAMPLE—Conservation of Momentum During Collision

A 2.35-kg ball (ball 1) is traveling at 5.50 m/s to the north. It glances off of a 2.75-kg ball (ball 2) that is at rest. After the collision, the first ball ends up traveling to the west at 1.00 m/s. (a) What is the final speed of the 2.75-kg ball? (b) Is this an elastic collision?

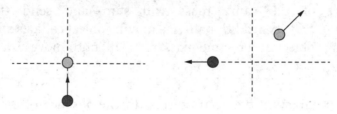

Solution:

(a) Note that there is no momentum in the *x*-direction prior to the collision. After the collision, the momentum of ball 1 to the left has to be equal to the *x*-component of the momentum of ball 2 to the right.

$$x\text{-direction:} \quad 0 = m_1 v_{1fx} + m_2 v_{2fx}$$

$$v_{2fx} = \frac{-m_1 v_{1fx}}{m_2}$$

$$v_{2fx} = -\frac{(2.35 \text{ kg})(-1.00 \text{ m/s})}{2.75 \text{ kg}} = 0.85 \text{ m/s}$$

In the *y*-direction, ball 1 has positive momentum prior to the collision, and ball 2 has no momentum. After the collision, ball 1 has no *y*-component to its momentum, so the *y*-component of the momentum of ball 2 after the collision has to be equal to the momentum of ball 1 before the collision:

$$y\text{-direction:} \qquad m_1 v_{1iy} = m_2 v_{2fy}$$

$$v_{2fy} = \frac{m_1 v_{1iy}}{m_2} = \frac{(2.35\,\text{kg})(5.50\,\text{m/s})}{2.75\,\text{kg}} = 4.70\,\text{m/s}$$

Now we apply the Pythagorean theorem to the *x*- and *y*-components of ball 2 to find its final velocity.

$$v_2 = \sqrt{v_{2fx}^2 + v_{2fy}^2} \qquad v_2 = \sqrt{(0.85\,\text{m/s})^2 + (4.70\,\text{m/s})^2} = 4.78\,\text{m/s}$$

(b) For the collision to be elastic, total kinetic energy before and after the collision must be the same. Kinetic energy is not a vector, so we do not calculate with components.

$$K_{before} = \frac{1}{2}m_1 v_{1i}^2 = \frac{1}{2}(2.35\,\text{kg})(5.5\,\text{m/s})^2 = 35.5\,\text{J}$$

$$K_{after} = \frac{1}{2}m_1 v_{1f}^2 + \frac{1}{2}m_2 v_{2f}^2 = \frac{1}{2}(2.35\,\text{kg})(1.00\,\text{m/s})^2 + \frac{1}{2}(2.75\,\text{kg})(4.78\,\text{m/s})^2 = 32.6\,\text{J}$$

The collision is not elastic; some kinetic energy was given up to thermal energy during the collision.

EXAMPLE—Totally Inelastic Collision

A 2,000-kg vehicle sitting at a stoplight is hit from behind by a 2,500-kg vehicle. The cars lock bumpers and slide into the intersection a distance of 5 m. If the coefficient of friction between the tires and pavement is 0.3, determine the speed of the larger car prior to the collision.

Solution:

This is a totally inelastic collision, so momentum is conserved but kinetic energy is not conserved.

We can use Newton's second law to determine the acceleration while the joined cars come to a stop. The acceleration is final velocity (which is zero) minus initial velocity. Then the friction force is equal to the normal force of the two cars on the surface times the coefficient of friction. So the friction force is the force in Newton's law, which is equal to μmg. Notice that mass cancels.

$$\mathbf{F}_{friction} = m\mathbf{a} \qquad v_f = 0$$
$$\mu mg = -m\mathbf{a}$$
$$-\mu g = \mathbf{a}$$
$$\mathbf{a} = -(0.3)(9.8\,\text{m/s}^2) = -2.94\,\text{m/s}^2$$
$$v_f^2 = v_o^2 - 2ad$$
$$0 = v_o^2 - (2)(-2.94\,\text{m/s}^2)(5\,\text{m})$$
$$v_o = \sqrt{29.4} = 5.4\,\text{m/s}$$

The initial velocity from this calculation is the velocity of the combined cars **after they** collide. Apply this to the conservation of momentum during the collision to find **the velocity** of the oncoming car. Remember that the first car was stationary.

$$m_1v_1 + m_2v_2 = (m_1 + m_2)v_o$$
$$0 + (2,500\text{ kg})v_2 = (4,500\text{ kg})(5.4\text{ m/s})$$
$$v_2 = 9.7\text{ m/s}$$

EXERCISE 6.3

1. On an air track, a 2.0-kg cart moving at 1.0 m/s to the right collides with a **1.0-kg cart** moving at 3.0 m/s to the left. After the collision, the 2.0-kg cart is moving at **2.0 m/s** to the left. The velocity of the 1.0-kg cart after the collision is

   A. 1 m/s to the right
   B. 1 m/s to the left
   C. 3 m/s to the right
   D. 2 m/s to the left

2. Object A, which is moving to the right at speed 2v, collides head-on and **totally inelastically** with an identical object B moving to the left at speed 4v. What occurs **after the collision?**

   A. Object A moves to the right at v and object B moves to the left at 2v.
   B. Object A moves to the right at 2v and object B moves to the left at v.
   C. The objects move together to the right at speed v.
   D. The objects move together to the left at speed v.

3. A cart rolls across a level floor and strikes a wall elastically at +10 m/s. On its **way back,** it strikes an identical cart rolling toward it at 20 m/s. They collide elastically. **What is a** possible result of that collision?

   A. One cart could be moving at 6 m/s and the other cart moving at −5 m/s.
   B. One cart could be moving at 4 m/s and the other cart moving at −7 m/s.
   C. One cart could be moving at +15 m/s and the other cart moving at −5 m/s.
   D. Both carts could be moving away from each other at a speed of 4 m/s.

4. A cart on a level air track is moving at +2 m/s when it strikes and connects **to a stationary** cart of the same mass. The final velocity of the connected carts is:

   A. +4 m/s
   B. +2 m/s
   C. +1 m/s
   D. −1 m/s

5. A coin sliding across a smooth surface at 0.5 m/s strikes a stationary identical **coin head**-on (or their center of masses aligned). The first coin stops at the time of **collision. What** happens to the coin that was stationary?

   A. The second coin remains stationary also.
   B. The second coin goes off at an angle to the original line of motion.
   C. The second coin continues on in a line in the same direction at 0.5 m/s.
   D. The second coin continues on in a line in the same direction at 1.0 m/s.

6. On an air track, a 0.5-kg cart moving at 1.0 m/s to the right collides with a 1.0-kg cart moving at 2.0 m/s to the left. After the collision, the 0.5-kg cart is moving at 2.0 m/s to the left. The velocity of the 1.0-kg cart after the collision is:

   A. 1 m/s to the right
   B. 1 m/s to the left
   C. 0.5 m/s to the right
   D. 0.5 m/s to the left

7. Object A, which is moving to the right at speed 4v, collides head-on and totally inelastically with an identical object B moving to the left at speed 2v. What occurs after the collision?

   A. Object A moves to the right at v and object B moves to the left at 2v.
   B. Object A moves to the right at 2v and object B moves to the left at v.
   C. The objects move together to the right at speed v.
   D. The objects move together to the left at speed v.

8. Which of the following are conserved in a totally inelastic collision of two objects, in the absence of external forces on the system of objects?

   A. Only momentum
   B. Only kinetic energy
   C. Both momentum and kinetic energy
   D. Neither momentum nor kinetic energy

9. Which of the following statements is always true for elastic collisions, assuming there is no external force applied in the direction of motion?

   A. Momentum is conserved but kinetic energy is not conserved.
   B. Kinetic energy is conserved but momentum is not conserved.
   C. Both momentum and kinetic energy are conserved.
   D. Energy is not conserved, but the objects exchange velocities during the collision.

10. During an inelastic collision, kinetic energy of the system is less after the collision than before the collision because:

   A. Kinetic energy is changed to other forms of energy.
   B. Kinetic energy has to decrease as momentum increases.
   C. Kinetic energy cannot be conserved during any collision.
   D. Kinetic energy decreases any time a force is applied to a system.

Oscillations and Waves

- Motion of Spring and Pendulum
- Simple Harmonic Motion
- Energy of Oscillators
- Properties of Waves
- Interference and Superposition
- Sound

This combination of topics, which might be summarized as "oscillations in a medium," can be very expansive, including the motions of springs and pendulums, mechanical waves in various materials, sound waves, sound intensity, wave interference, resonance, harmonics, and music. The selection of topics here is limited—including the common areas covered in a first course in physics, with example problems and solutions covering problems that most commonly offer difficulty for students.

Motion of Spring and Pendulum

Harmonic motion: Motion that repeats periodically, such as the oscillation of a pendulum or a mass on a spring.

Period, T: Time for one complete oscillation, usually measured in seconds.

Frequency, f: Number of oscillations per second, measured in cycles/sec or s^{-1} or hertz (Hz).

Amplitude: Maximum displacement from the equilibrium position. Amplitude is proportional to the energy of the oscillator or wave. An oscillator may decrease in amplitude but maintain a constant frequency.

Damping: Decrease in amplitude or decrease in energy due to energy loss during oscillation (usually friction).

Pendulum: Mass attached at the end of a string or chain allowed to oscillate under the influence of gravity. Period is proportional to square root of length and inversely proportional to square root of g.

$$T = 2\pi\sqrt{\frac{L}{g}}$$

Spring motion: Oscillation of an object on a spring. The period of oscillation of a spring is directly proportional to the square root of the mass on the spring and inversely proportional to the square root of the **spring constant k.** The spring

NGSS PS4-1 and PS4-3

constant describes the "stiffness" of the spring, or the amount of force needed to stretch or compress the spring by a certain length.

Hooke's Law: For a mass oscillating on a spring, the force applied by the spring $\bar{F}$ is proportional to the displacement $\bar{x}$ of the spring from its equilibrium position—and the force is in the opposite direction of the displacement: $\bar{F} = -k\bar{x}$.

The **total energy** of an oscillating spring is $\frac{1}{2}kA^2$, where k is the **elastic constant or spring constant** and A is the **amplitude**. In the absence of friction, the total energy remains constant as a spring oscillates, and the total energy is the sum of the potential energy and kinetic energy at any distance from equilibrium, x:

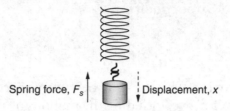

Spring force, F_s Displacement, x

EXAMPLE—Force versus Displacement Graph of Spring

Use the Force versus Displacement graph below to answer the following questions about a mass of 500 g attached to the spring. (a) What is the spring constant of the spring? (b) Calculate the period of the spring-mass system when it oscillates. (c) Determine the frequency of oscillation of the spring-mass system. (d) What is the work required to displace the spring from its equilibrium position to $x = 0.005$ m? (e) What is the work required to displace the spring from $x = 0.005$ m to $x = 0.01$ m?

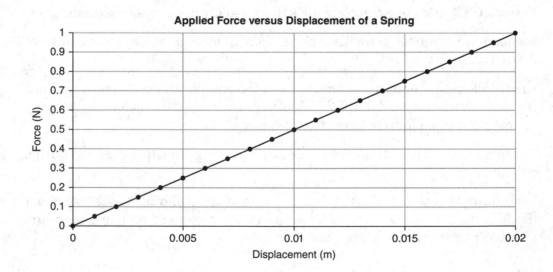

Solution:

(a) $F = kx$ for the force applied to displace a mass on a spring. Therefore, the slope of this plot is the spring constant for the spring.

$$\text{slope} = \frac{\Delta F}{\Delta x} = \frac{1\,\text{N}}{0.02\,\text{m}} = 59\,\text{N/m}$$

(b) Since we know the spring constant and mass, the period can be calculated:

$$T = 2\pi \sqrt{\frac{m}{k}} = 2\pi \sqrt{\frac{0.5 \text{ kg}}{50 \text{ N/m}}} = 0.63 \text{ s}$$

(c) Frequency is reciprocal of period: $f = 1.6$ Hz.

(d) Work is the change in potential energy, so this can be determined in two ways:

   (1) Use the equation $U = \frac{1}{2} k(\Delta x)^2$

   $$W = \Delta U = \frac{1}{2} (50 \text{ N/m})(0.005 \text{ m} - 0)^2 = 0.00063 \text{ J}$$

   (2) Find the area from $x = 0$ to $x = 0.005$ m from the graph.

   $$\text{Area} = \frac{1}{2} (0.005 \text{ m})(0.25 \text{ N}) = 0.00063 \text{ J}$$

(e) Use similar methods as above, but remember that we're not stretching the spring from equilibrium. We're stretching it the same distance as before, but the spring force is larger in this case, so we have to consider the *change* in potential energy for method (1).

   (1) $W = \Delta U = U_f - U_o = \frac{1}{2} (50 \text{ N/m})(0.01 \text{ m})^2 - \frac{1}{2} (50 \text{ N/m})(0.005)^2 = 0.0019 \text{ J}$

   (2) Area (trapezoid from 0.005 to 0.01) $= \frac{1}{2} (0.25 \text{ N} + 0.5 \text{ N})(0.005 \text{ m}) = 0.0019 \text{ J}$

   In summary, the method of taking area from the graph may be easier.

EXAMPLE (Advanced)—Harmonic Motion Graph and Equation

The graph below describes the harmonic motion of a 10-kg mass oscillating vertically on a spring.

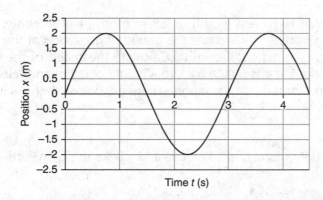

(a) What is the period of the motion? (b) What is the frequency of the motion? (c) What is the amplitude of the motion? (d) What is the spring constant of the spring? (e) Write the equation for the above graph.

Solution:

(a) Period is the time between adjacent crests or troughs, which is 10 s.

(b) Frequency, f, is the reciprocal of period, so $f = 0.1$ s.

(c) The amplitude is maximum displacement from equilibrium, which is 3 m.

(d) Using the spring equation:

$$T = 2\pi\sqrt{\frac{m}{k}}$$

$$10\ s = 2\pi\sqrt{\frac{10\ kg}{k}}$$

$$k = 3.9\ N/m$$

(d) The graph is a sine function (since amplitude is zero when $t = 0$). The equation is of the form:

$$x(t) = A\ \sin(2\pi ft)$$

so:

$$x(t) = (3\ m)\sin(0.2\pi t)$$

EXAMPLE—Energy of a Pendulum

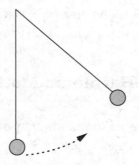

A pendulum is measured to have a length of 2.3 m (from pivot to center of mass) and has a mass of 2.5 kg hanging from it. (a) Calculate the period of oscillation of the pendulum on Earth. (b) What is the frequency of its oscillation? (c) Calculate the speed of the pendulum bob as it passes through equilibrium, if it is pulled back to a height of 0.2 m and released. (d) What is the period of the pendulum's motion on the moon, where g is one-sixth the value here on Earth?

Solution:

(a) The length of the pendulum, L, is 2.3 meters and g is 9.8 m/s/s on Earth.

(b) $f = 1/T = 0.33$ Hz

(c) Convert PE to KE and solve for v:

$$\Delta PE = \Delta KE$$

$$mg\Delta h = \tfrac{1}{2}\ mv^2$$

mass cancels:

$$(9.8\ m/s^2)(0.2\ m) = \tfrac{1}{2}\ v^2$$

$$v^2 = 3.92$$

$$v = 1.98\ m/s$$

(d) $\quad T = 2\pi\sqrt{\dfrac{L}{(g/6)}} = 2\pi\sqrt{\dfrac{2.3\ m}{1.63\ m/s^2}} = 7.42\ s$

The pendulum oscillates more slowly when g is lower; i.e., each oscillation takes longer. And the pendulum won't oscillate if there is no gravitational field.

EXAMPLE—Hooke's Law and Motion of a Spring

Use the "Force versus Displacement graph below to answer the following questions about a mass of 500 g attached to a spring. (a) What is the spring constant of the spring? (b) Calculate the period of the spring-mass system when it oscillates. (c) Determine the frequency of oscillation of the spring-mass system. (d) What is the work required to displace the spring from its equilibrium position to $x = 0.005$ m? (e) What is the work required to displace the spring from $x = 0.005$ m to $x = 0.01$ m?

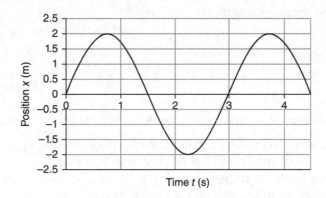

Solution:

(a) $F = kx$ for the force applied to displace a mass on a spring. Therefore, the slope of this plot is the spring constant for the spring.

$$\text{slope} = \frac{\Delta F}{\Delta x} = \frac{1\ N}{0.02\ m} = 50\ N/m$$

(b) Since we know the spring constant and mass, the period can be calculated:

$$T = 2\pi\sqrt{\frac{m}{k}} = 2\pi\sqrt{\frac{0.5\ kg}{50\ N/m}} = 0.63\ s$$

(c) Frequency is reciprocal of period: $f = 1.6$ Hz

(d) Work is the change in potential energy, so this can be determined in two ways:

(1) Use the equation $U = \frac{1}{2}\,k(\Delta x)^2$

$$W = \Delta U = \frac{1}{2}\,(50\ N/m)(0.005\ m - 0)^2 = 0.00063\ J$$

(2) Find the area from $x = 0$ to $x = 0.005$ m from the graph.

$$\text{Area} = \frac{1}{2}\,(0.005\ m)(0.25\ N) = 0.00063\ J$$

(e) Use similar methods as above, but remember that we're not stretching the spring from equilibrium. We're stretching it the same distance as before, but the spring force is larger in this case, so we have to consider the *change* in potential energy for method (1).

(1) $W = \Delta U = U_f - U_o = \frac{1}{2}$ (50 N/m)(0.01 m)^2 – $\frac{1}{2}$ (50 N/m)(0.005)^2 = 0.0019 J

(2) Area (trapezoid from 0.005 to 0.01) = $\frac{1}{2}$ (0.25 N + 0.5 N)(0.005 m) = 0.0019 J

In summary, the method of taking area from the graph may be easier.

EXERCISE
7.1

1. A mass attached to a spring is displaced by 4 cm and released to set it into oscillation with a period of time T. Then the mass is displaced by 8 cm and set into oscillation again. What is the period of the second oscillation?

   A. T
   B. $T/2$
   C. $2T$
   D. $4T$

2. The period of oscillation of a mass-spring system is affected by:

   A. the mass attached to the spring but not the length of the spring
   B. the elastic constant of the spring but not the attached mass
   C. the length of the spring and the value of g
   D. the spring constant of the spring and the value of g

3. If the length of a simple pendulum is quadrupled, the period of its oscillation will be:

   A. unaffected
   B. cut in half
   C. doubled
   D. quadrupled

4. A swinging pendulum can be considered a simple harmonic oscillator only if:

   A. the pendulum length is large
   B. the mass on the pendulum is small
   C. the angle of release is small
   D. the value of g is small

5. A mass of 100 g is connected to a hanging spring and the mass-spring system comes to equilibrium. Another 100-g mass is added and the spring stretches by 10 cm. What is the elastic constant of the spring?

   A. 100 N/m
   B. 9.8 N/m
   C. 10 N/m
   D. 19.6 N/m

Simple Harmonic Motion

Simple Harmonic Motion: During simple harmonic motion, a restoring force is exerted on the oscillator proportional to the displacement of the oscillator from its equilibrium position. The restoring force is in the opposite direction of the displacement. Springs are examples of simple harmonic motion, since the spring equation is $F = -k\Delta x$, where the negative sign indicates the opposite directions of the force and displacement.

Pendulums, on the other hand, approximate simple harmonic oscillators but are not exactly harmonic. The displacement of a pendulum is a sine function, which is not linear (and therefore not directly proportional). The pendulum approximates SHM as long as the angle of displacement is about 15° or less.

The equation to describe simple harmonic motion for position of the oscillator as a function of time is:

$$x(t) = A \cos 2\pi f t$$

where x is the position of the oscillator (in meters), A is the amplitude (in meters), f is the frequency (in s^{-1} or Hz), and t is the time. The motion is the same whether sine or consider is used, except for the position of the oscillator at $t = 0$. When the motion starts at its amplitude at $t = 0$, such as pulling down a mass on a spring to start it, the sine function is used. When the motion starts at the equilibrium position at $t = 0$, the cosine function is used.

EXERCISE 7.2

Questions 1–4 refer to this equation.

The equation of a simple harmonic oscillator is: $x = (1.5 \text{ m}) \sin 18.8\, t$.

1. What is the frequency of the oscillator?

   A. 6 Hz
   B. 3 Hz
   C. 2 Hz
   D. 18.8 Hz

2. What is the amplitude of the oscillator?
   A. 18.8 m
   B. 1.5 m
   C. 3.0 m
   D. 4.5 m

3. What is the position of the oscillator when $t = 0$?

   A. 0
   B. 1.5 m
   C. 3.0 m
   D. 18.8 m

4. Instead of starting the motion of the oscillator by giving it a push when it is at equilibrium, the oscillator starts its motion when it is pulled back and released. How will this change the oscillation?

   A. the frequency will increase.
   B. the frequency will decrease.
   C. the elastic constant will change.
   D. period and frequency will be the same, but position at $t = 0$ will be different.

5. The oscillator is started with twice the amplitude. How does its frequency change?

   A. the frequency will increase.
   B. the frequency will decrease.
   C. the frequency will stay the same.

Energy of Oscillators

The energy of mechanical oscillators such as spring and pendulums is in the form of kinetic energy or potential energy or both. The oscillator is usually changing in position from maximum KE and minimum PE to a point where it has maximum PE and minimum KE.

The total energy of an oscillating spring is $\frac{1}{2} kA^2$, where k is the elastic constant or spring constant and A is the amplitude. In the absence of friction, the total energy remains constant as a spring oscillates, and the total energy is the sum of the potential energy and kinetic energy at any distance from equilibrium, x:

$$E_T = \frac{1}{2} mv^2 + \frac{1}{2} kx^2$$

EXAMPLE—Graph of Harmonic Motion

Describe the motion that this graph illustrates if it depicts position in meters as a function of time in seconds for a spring oscillator with a 2-kg attached mass.

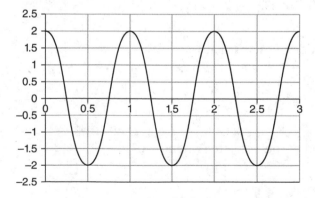

Solution:

The amplitude is 2 m, which is the maximum displacement from zero—either positive or negative. The period is 1 s, which is the time from a point on a wave to the same point on the next wave.

The frequency is 1 Hz, since $f = 1/T$.

When we substitute these values into the simple harmonic motion equation, we get the following equation for the motion:

$$x = (2\ m)\cos 6.28\ t$$

This information can be used to find the spring constant, k, of the spring:

$$T = 2\pi\sqrt{\frac{m}{k}}$$

$$1s = 2\pi\sqrt{\frac{2\ kg}{k}}$$

$$1 = (4\pi^2)\left(\frac{2}{k}\right)$$

$$k = 79\ N/m$$

EXAMPLE (Advanced)—Equation for Harmonic Motion

A spring with a mass of 2.0 kg attached to it undergoes simple harmonic motion of the form:

$$x(t) = 0.300\cos 7.50t$$

(where x is in meters, t is in seconds, and $7.50t$ is in radians).

(a) What is the amplitude of the motion? (b) What is the frequency (f) of the system? (c) What is the position of the oscillator at time $t = 2$ s? (d) Calculate the total energy of the spring-mass system as it oscillates.

Solution:

(a) The amplitude is 0.300 m, since the equation is of the form $x(t) = A\cos(2\pi ft)$.

(b) Compare the equation form to the equation: $7.50 = 2\pi f$

$$f = 1.2\ Hz$$

(c) Substitute 2 s into the equation, using the values of A and f just determined:

$$x(t) = (0.3\ m)\cos[(2)(\pi)(1.2\ Hz)(2\ s)] = 1.02\ m$$

(d) The energy is periodically converted from potential energy (at amplitude of oscillation) to kinetic energy (at equilibrium)—but the total mechanical energy remains constant. We can use the amplitude (where $x = A$) to find the maximum when the energy is entirely potential:

$$U_{max} = \tfrac{1}{2}\,kA^2$$

First, determine k by using the spring equation:

$$T = \frac{1}{f} = 2\pi\sqrt{\frac{m}{k}}$$

$$\frac{1}{1.2 \text{ Hz}} = 2\pi\sqrt{\frac{2 \text{ kg}}{k}}$$

$$k = 114 \text{ N/m}$$

Now calculate the maximum energy: $U = \frac{1}{2}kA^2 = \frac{1}{2}(114 \text{ N/m})(0.3 \text{ m})^2 = 5.1 \text{ J}$

EXERCISE 7.3

Questions 1–2 use this graph.

1. The graph below shows data for amplitude as a function of time for an **object with** a mass of 0.5 kg oscillating on a spring. At which of the following times is the **spring potential** energy a maximum?

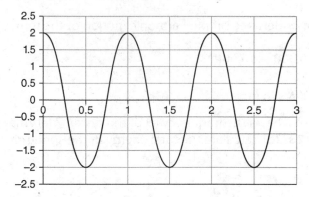

A. $t = 0.25$ s
B. $t = 0.50$ s
C. $t = 1.75$ s
D. $t = 2.75$ s

2. When is the kinetic energy of the spring maximum?

A. $t = 0$ s
B. $t = 0.50$ s
C. $t = 1.75$ s
D. $t = 2.0$ s

3. A spring is placed on a horizontal, frictionless surface and compressed **a distance of** 0.10 m from its rest position with a ball of mass 0.10 kg placed at its end. When **the spring is** released, the ball leaves the spring traveling at 10 m/s. What is the spring **constant?**

A. 50 N/m
B. 100 N/m
C. 500 N/m
D. 1,000 N/m

4. As a simple harmonic oscillator decreases in amplitude under the influence of friction:

   A. Period and frequency both decrease
   B. Total mechanical energy and frequency both decrease
   C. Total mechanical energy decreases and period increases
   D. Total mechanical energy decreases and frequency remains constant

5. An object attached to a spring is compressed a distance of 10 cm and released. Which statement is true about the spring force on the object as it oscillates? (Assume no friction.)

   A. The spring force is maximum when the object moves through equilibrium.
   B. The spring force is maximum each time the object is 10 cm from equilibrium.
   C. The spring force is maximum each time the object is 5 cm from equilibrium.
   D. The spring force is maximum each time the object reaches the point from which it was released and minimum when it reaches a point 10 cm on the other side of equilibrium.

Properties of Waves

Wave: Traveling disturbance that transmits energy from one place to another.

The model below illustrates how the oscillations of molecules in a **transverse** wave are perpendicular to the direction in which the wave travels (or is propagated).

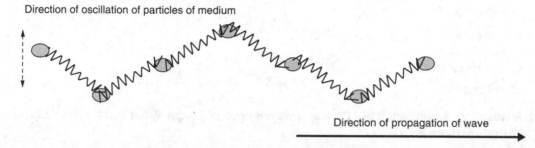

Direction of oscillation of particles of medium

Direction of propagation of wave

The model below shows how the oscillations of a **longitudinal** wave are along the same direction as the propagation of the wave. Sound travels as longitudinal oscillations.

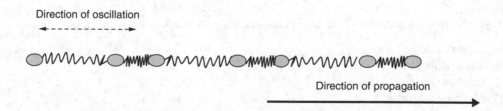

Direction of oscillation

Direction of propagation

Wavelength (λ): Distance along the direction the wave is moving between points of the same phase on adjacent waves.

Wave speed: The product of frequency and wavelength $v = f\lambda$.

Amplitude (A): The maximum displacement of an oscillator from its equilibrium position.

Natural Frequency: Frequency at which a system will oscillate when free of external disturbances (such as damping or applied forces).

Resonance: Response to a driving force of the same natural frequency. For example, buildings from four to fourteen stories tall are resonant with earthquake vibrations.

Equilibrium: The position of a mechanical oscillator where there is no net force and no acceleration. An example of this is an object hanging at rest on a stretched spring.

> Remember that the frequency and period of a harmonic oscillator do not change as the oscillator loses amplitude (or energy).

EXERCISE 7.4

1. When a mechanical wave travels from a "faster" medium (i.e., wave speed is greater) to a "slower medium":

   A. wavelength and frequency decrease
   B. wavelength and frequency increase
   C. wavelength increases and frequency decreases
   D. wavelength decreases and frequency does not change

2. A certain wave is produced by the oscillations of particles of a medium in the positive y- and negative y-direction, as the wave pulses travel in the positive x-direction. What type of wave is this?

   A. Transverse
   B. Longitudinal
   C. Compressional
   D. Spherical

3. A wave has a frequency of 10 Hz and wavelength of 20 cm. What is the speed of the wave in that medium?

   A. 2 m/s
   B. 200 m/s
   C. 0.5 ms
   D. 1 m/s

4. A sound has a frequency of 5,000 Hz. What type of wave is it?

   A. supersonic
   B. longitudinal
   C. transverse
   D. electromagnetic

5. A sound wave, with speed of about 340 m/s in air, has a frequency of 5,000 Hz. What is the wavelength of this sound?

   A. 6.8 cm
   B. 0.68 m
   C. 14.7 m
   D. 14.7 cm

Interference and Superposition

Phase: Point on a wave where an oscillating element has a specific amplitude and direction of movement. Phase varies sinusoidally as a function of time for a wave. When waves meet or interfere in phase, i.e., crest to crest or trough to trough, they **constructively interfere.** When waves meet out of phase, they **destructively interfere.**

Standing Waves: Produced when a wave reflects from a boundary and interferes with itself. Where crests meet crests, constructive interference occurs and the wave amplitude increases. Where crests and troughs meet, destructive interference occurs and the waves may cancel or decrease in amplitude. Standing waves produced in tubes open at both ends and strings fixed at both ends have a wavelength that is two times the length. Standing waves produced in tubes that are closed at one end produce a wavelength that is four times the length.

In the diagram below, a wave pulse reflects from an open boundary and reflects back in the same phase and may reinforce other oncoming pulses.

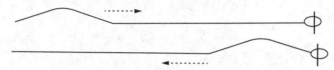

In the diagram below, a wave pulse reflects from a closed boundary and changes phase upon reflection.

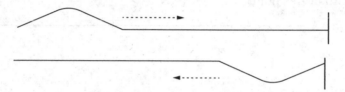

EXAMPLE—Wave Superposition

In the diagram below, wave fronts of the same frequency and wavelength are being produced by two sound sources. The sound waves travel outward in every direction from each source. Assume the waves are being produced in phase with each other at the source points.

At point E, which is the same distance from each source, we can assume that the waves meet each other in the same phase, so they *constructively reinforce* each other. At point E, we would hear the loudest sound with the largest amplitude and intensity.

At point B, however, the distances from the sources are different. Suppose the distance from one source is exactly 5 wavelengths and the distance from the other source is 6½ wavelengths. In that case, the waves would meet out of phase with each other and would *destructively reinforce*, or cancel, each other. A person at B might hear a very weak sound or no sound at all.

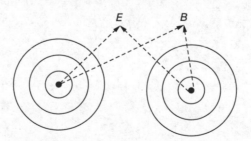

In the diagram below, two speakers are producing sounds of the same wavelength that are in phase with each other when they are produced. Point *A* is 4.5 m from the sound source on the left and 6.0 m away from the sound source on the right. Point *B* is 7.5 m away from the sound source in the left and 2.25 m away from the sound source on the right.

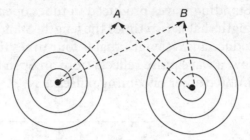

1. If the wavelength is 1.5 m, where will you hear the best sound? Explain.

2. Consider a point exactly halfway between the two speakers, if they are 6 m apart. Will a sound be heard at this point?

Sound

Sound: Longitudinal waves produced by oscillations of particles in a medium. Thus, sound cannot travel in a vacuum.

Harmonics: Set of frequencies for standing waves in a fixed space, such as a tube or string.

(1) For a tube open at both ends or a string fixed at both ends, the frequency for a harmonic, where *n* is the number of the harmonic and *v* is the speed of the wave:

$$\lambda_n = \frac{2L}{n}$$

where *n* is all whole numbers and *L* is length. Also:

$$f_n = \frac{nv}{2L}$$

(2) For a tube closed at one end (or a string fixed only at one end), the frequency for a harmonic, where n is the number of the harmonic and v is the speed of the wave:

$$\lambda_n = \frac{4L}{n}$$

where n is all odd numbers and L is length. Also:

$$f_n = \frac{nv}{4L}$$

Intensity, I: Power divided by area in watts per square meter. For sound, this is a measure of what we call "loudness." Sound intensity follows an inverse square law; i.e., doubling the distance from a sound source causes the intensity to decrease to one-fourth, and tripling the distance decreases intensity to one-ninth, and so on.

Intensity level: A scale to measure intensity, measured in decibels, where an intensity level or decibel level of zero is the lowest level humans can hear. A decibel level of about 60 dB is quiet room conversation, a decibel level of 110 dB produces hearing damage, and a decibel level of 120 dB produces pain.

Audible Frequency: Humans hear sounds of frequency from about 20 Hz to about 20,000 Hz. Sounds of higher frequency are called **ultrasonic** and frequencies below audible frequency are called **infrasonic.**

Beats: Pulsations of sound detected when two sounds with almost the same frequency are produced simultaneously so that they interfere. The beat frequency is equal to the difference of the two interfering frequencies.

Speed of Sound: Depends on density and elasticity of the medium. The speed of sound in dry air varies with temperature:

$$v_{air} = 331 \text{ m/s} + 0.6 \text{ T°C}$$

Unless other information is provided, use 340 m/s for speed of sound in air.

Objects that travel faster than the speed of sound are called **supersonic** and slower than the speed of sound is **subsonic.**

The speed of sound on a string increases with the tension in the string and decreases proportionally to the linear density (thickness) of the string.

Doppler Effect: As a source of sound and the person hearing the sound approach each other, the wavelengths "compress" and the person hears a higher frequency. As the source and/or receiver move away from each other, the sound appears to lower in frequency. For example, as a firetruck siren passes you and drives away, the **pitch** seems to become lower—and, of course, so does the loudness.

EXAMPLE—Harmonics in a Tube

If the length of a tube that is open at both ends is 2.0 m and the air oscillating in the tube is at 30°C, find: (a) the speed of sound in the air; (b) the wavelength of the first harmonic; (c) the frequency of the first harmonic (fundamental); (d) the frequency of the second harmonic (first overtone).

Solution:

(a) Speed of sound in air is: $v = (331 + 0.6TC°)$ m/s

$$v = 331 + 0.6(30) = 349 \text{ m/s}$$

(b) For an open tube: $\lambda_1 = 2L = 4.0$ m

(c) and (d) For an open tube:

$$f_n = \frac{nv}{2L}$$

$$f_1 = \frac{349 \text{ m/s}}{4.0 \text{ m}} = 87 \text{ Hz}$$

$$f_2 = \frac{(2)(349 \text{ m/s})}{4.0 \text{ m}} = 175 \text{ Hz}$$

EXAMPLE—Wave Superposition and Beats

The plot below was made by superimposing one Position versus Time graph on top of another Position versus Time graph. They start at the same time, but their frequencies and wavelengths are slightly different. You will notice that there is a regular pattern of places where the two overlap and reinforce, and there are places where they cancel. This happens about four times per second, so the "beat frequency" is 4 Hz and the two frequencies are 4 Hz apart.

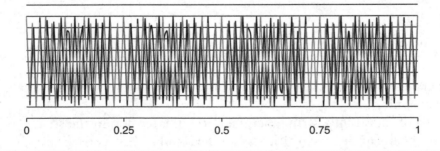

EXAMPLE—Harmonics in Open Tube

If the length of a tube that is open at both ends is 2.0 m and the air oscillating in the tube is at 30°C, find: (a) the speed of sound in the air; (b) the wavelength of the first harmonic; (c) the frequency of the first harmonic (fundamental); (d) the frequency of the second harmonic (first overtone).

Solution:

(a) Speed of sound in air is: $v = (331 + 0.6TC°)$ m/s

$$v = 331 + 0.6(30) = 349 \text{ m/s}$$

(b) For an open tube: $\lambda_1 = 2L = 4.0$ m

(c) and (d) For an open tube:

$$f_n = \frac{nv}{2L}$$

$$f_1 = \frac{349 \text{ m/s}}{4.0 \text{ m}} = 87 \text{ Hz}$$

$$f_2 = \frac{(2)(349 \text{ m/s})}{4.0 \text{ m}} = 175 \text{ Hz}$$

EXERCISE
7.6

1. Two tuning forks are sounded together, and we hear 10 beats per second. The shorter one is unmarked, and the longer one is marked 220 Hz. How can we determine the frequency of the unmarked tube?

   A. (220 + 10) Hz
   B. (220 − 10) Hz
   C. (220)(10) Hz
   D. (220/10) Hz

2. Two open plastic pipes produce the same 260 Hz frequency when they are struck. One pipe is sanded so that it is slight shorter, and the pipes produce 4 beats per second when they are struck at the same time. What is the frequency of the shorter pipe?

   A. 256 Hz
   B. 260 Hz
   C. 264 Hz
   D. 1,040 Hz

3. As a musician "warms up" a wind instrument ready to play, the temperature of the air in the instrument increases, resulting in these changes in the fundamental note the instrument will play:

A. an increase in frequency and wavelength
B. an increase in frequency and decrease in wavelength
C. an increase in wave speed and frequency
D. an increase in wave speed and decrease in frequency

4. A speaker is moved twice as far away from you. What is the resulting intensity of the sound where you sit, compared to what it was before the speaker was moved?

A. half the intensity
B. one-fourth the intensity
C. twice the intensity
D. four times the intensity

5. A tube open at both ends that is 1.5 m long is used to produce a sound by forcing a large blast of air across the end. What are the approximate wavelength and frequency of the lowest note produced?

A. Wavelength is 1.5 m and frequency is about 230 Hz.
B. Wavelength is 3.0 m and frequency is about 230 Hz.
C. Wavelength is 3.0 m and frequency is about 110 Hz.
D. Wavelength is 6.0 m and frequency is about 57 Hz.

Torque and Rotational Motion

·8·

- Rotational Kinematics
- Torque and Angular Acceleration
- Conservation of Angular Momentum

Rotational Kinematics

Angular Displacement, θ: Angle turned by a rotating object, measured in cycles, turns, rotations, degrees, or radians. One complete cycle is equal to 360° or 2π radians. Angular displacement in radians is equal to linear displacement divided by radius:

$$\Delta x = \theta R$$

Angular velocity, ω: Rate at which an object is rotating in units such as radians per second or rotations per minute (rpm). Angular velocity is a vector, with direction determined by a right-hand rule. Angular velocity (in radians per second) is equal to linear velocity v divided by R:

$$\omega_{ave} = \frac{\Delta\theta}{\Delta t} = \frac{v}{R}$$

$$\omega_f^2 = \omega_i^2 + 2\alpha\theta$$

$$\theta = \omega_i t + \tfrac{1}{2}\alpha t^2$$

The angular velocity of the Earth on its axis is counterclockwise as you look downward on the geographic North Pole. The rate at which the Earth turns is 2π radians per day, and that angular velocity is the same for every point on the planet.

> The equations describing rotational motion are analogous to the kinematic equations for linear motion, as careful inspection of the above equations will show.

NGSS HS-PS2-4 and HS-ESS1-4

Angular Acceleration, α: Rate of change in angular velocity. The angular acceleration is a vector that is in the same direction as the net torque. Angular acceleration is equal to linear acceleration **a** in rad/s/s, divided by the radius:

$$\alpha = \frac{\Delta\omega}{\Delta t} = \frac{a}{R}$$

Period of Rotation, T: Time for one complete rotation of an object or system about a defined axis. Period (in seconds) is the reciprocal of frequency (measured in s^{-1} or hertz).

$$T = \frac{1}{f}$$

Rotational Inertia, I: A property of an object or system that describes how difficult it is to start spinning an object or to stop it from spinning. It is equal to a constant, k, that describes the distribution of mass within the object times the mass times the radius squared. For example, the rotational inertia of a solid sphere is $\frac{2}{5}mr^2$, and the rotational inertia of a hollow sphere is $\frac{2}{3}mr^2$:

$$I = kmr^2$$

Rotational Kinetic Energy, K or KE: The kinetic energy of a rotating object; has the same symbol and units as linear kinetic energy. If an object has both rotational and translational kinetic energy, such as a ball rolling across a surface, the total kinetic energy is the sum of both.

$$KE_R = \tfrac{1}{2}I\omega^2$$
$$KE_{total} = \tfrac{1}{2}mv^2 + \tfrac{1}{2}I\omega^2$$

> An object has rotational inertia whether it is rotating or not. It is a property of the object regardless of its state of motion. The rotational inertia depends on where the axis of rotation is set, so changing the position of the axis of rotation will change the rotational inertia.

EXERCISE 8.1

1. A wheel has a force exerted on it so that it begins to rotate faster and faster clockwise. Which of the following is <u>not</u> also clockwise?

   A. Angular velocity
   B. Angular momentum
   C. Angular acceleration
   D. Moment of inertia

2. A ball with mass 1.0 kg and radius 0.2 m has moment of inertia equal to $\frac{2}{5}mr^2$. If the ball is rolling across the floor with linear speed is 2 m/s, determine the total kinetic energy of the ball. The angular velocity is equal to the linear speed divided by the radius.

   A. 4.0 J
   B. 3.2 J
   C. 2.8 J
   D. 0.5 J

3. When a turntable with a radius of 8.0 cm is rotating at 45 rev/min, what is the linear speed of a point on the outside rim of the turntable?

   A. 22.6 m/s
   B. 5.6 m/s
   C. 3.6 m/s
   D. 0.38 m/s

4. If you walk ¼ of the way around a circular track that has a radius of 20 m, what are the linear distance and angular distance you have moved?

   A. 15.7 meters, $\pi/4$ radians
   B. 62.8 meters, $\pi/2$ radians
   C. 31.4 meters, $\pi/4$ radians
   D. 31.4 meters, $\pi/2$ radians

5. You walk in an arc around a circular track with a radius of 20 m and cover an angular displacement of π radians. What is the linear distance, in meters, that you have walked?

   A. 62.8
   B. 31.4
   C. 125.7
   D. 15.7

6. A student swings his arm in a vertical circle and is able to make 5 complete swings in 1.25 s. What is the average angular velocity for the student's arm swinging in a circle?

   A. About 4 radians/second
   B. About 25 radians/second
   C. About 31 radians/second
   D. About 6.28 radians/second

7. A bicycle wheel has a small orange ball attached to the spokes halfway to the center from the rim and a small yellow attached to a spoke closer to the outside rim. When the wheel is spinning, compare the angular velocity and the linear velocity of the two balls.

   A. The balls both have the same angular velocity, but the yellow ball has a larger linear velocity.
   B. The balls both have the same angular velocity and the same linear velocity, since they are attached to the same spoke.
   C. The balls both have the same linear velocity, but the yellow ball has a larger angular velocity.
   D. The balls both have the same angular velocity, but the orange ball has a larger linear velocity.

8. A child rides a horse on a carousel that is rotating at a constant angular velocity of 2 cycles per minute, which is 4π radians every 60 s, or $\pi/15$ radians per second. If the child and horse are located 3 m from the center, what is the linear speed of the child and horse (in m/s)?

   A. $v = \omega R = 12\pi$ m/s
   B. $v = \omega R = 6\pi$ m/s
   C. $v = \omega R = \pi/5$ m/s
   D. $v = \omega/R = \pi/45$ m/s

9. Two students are running a relay race for one lap around a circular track. Student 1 runs halfway around the track and hands the baton to Student 2, who runs the second half in less time. Which is a true statement about the students' race?

    A. The students' angular displacements are the same.
    B. The students' average linear speeds are the same.
    C. The students' average angular velocities are the same.
    D. Student 1 has a larger angular velocity than Student 2.

Torque and Angular Acceleration

Torque, τ: External force exerted on an object or system multiplied by the distance from the line of force to the axis of rotation. It is a vector quantity with direction determined by a right-hand rule. Torque can also be determined by multiplying rotational inertia (I) and angular acceleration (α).

$$\tau = r_{\perp}F = I\alpha$$

In the case below, the pivot, or axis of rotation, is at the center of the board, so the weight of the board exerts no torque. The force F is exerted at a distance L from the axis of rotation, which will cause the board to begin to rotate clockwise.

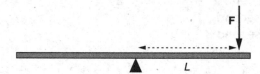

The **right-hand rule** can be used to determine the direction of torque. As shown below, point your index finger outward from the axis of rotation to the line of force (x-direction on the diagram). Then point the other fingers of the right hand along the direction of the force (y-direction). The outstretched thumb is the direction of torque (z-direction).

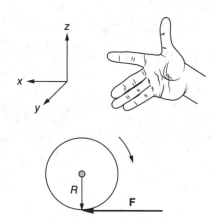

Another way to describe torque and rotational motion is to curl the fingers of your right hand around in the direction the object is rotating. Your thumb, then, points in the direction of the torque that causes the rotation and the directions of both the **angular velocity** and

angular momentum. Using either method, the torque produced by the force *F* applied in this way is directed into the page, causing the wheel to turn clockwise.

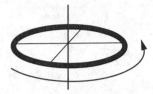

In this diagram, a wheel is rotating into and out of the page around an axis that is up and down on the page. You curl your fingers counterclockwise as you look down on the wheel. Your thumb points along the axle upward on the page—which is the direction of torque, angular velocity, and angular momentum in this case.

> Note: In most introductory courses, the description clockwise or counterclockwise will be sufficient, and the right-hand rule can be reserved for a more advanced course.

Equilibrium: A situation where the net force on an object or system of objects is zero and the net torque on the object or system is zero. An object or system at equilibrium has no linear or angular acceleration. Net torque is zero when the sum of clockwise torques equals the sum of counter-clockwise torques.

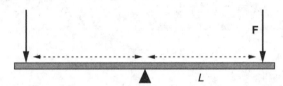

In this case, the **net torque** now adds to zero, and the board does not rotate. The board is in **rotational equilibrium.**

EXAMPLE—Rotating Platform and Torque

A merry-go-round with a mass of 40 kg, a radius of 2.0 m, and moment of inertia of 80 kg·m² starts at rest and reaches an angular speed of 10 rpm (revolutions per minute) in 20 s. Calculate the amount of force (tangential to the outside edge) that will be required to produce this angular acceleration.

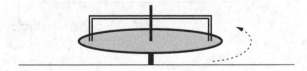

Solution:

Step 1. Change 10 rpm to radians per second.

$$\left(\frac{10 \text{ rev}}{1 \text{ min}}\right)\left(\frac{1 \text{ min}}{60 \text{ s}}\right)\left(\frac{2\pi \text{ rad}}{1 \text{ rev}}\right) = 1.05 \text{ rad/s}$$

Step 2. Outline the variables.

$\omega_i = 0$

$\omega_f = 1.05$ rad/s

$m = 40$ kg

$R = 2.0$ m

$I = 80$ kg $\times$ m^2

$t = 20$ s

Step 3. Substitute into appropriate equations.

$$\alpha = \frac{\Delta\omega}{\Delta t} = \frac{\omega_f - \omega_i}{t} = \frac{1.05 \text{ rad/s} - 0}{20 \text{ s}} = 0.053 \text{ rad/s}^2$$

$$\tau = I\alpha = (80 \text{ kg} \times \text{m}^2)(0.053 \text{ rad/s}^2) = 4.2 \text{ N} \times \text{m}$$

$$\tau = R_\perp \mathbf{F}$$

$$\mathbf{F} = \frac{\tau}{R} = \frac{4.2 \text{ N} \times \text{m}}{2.0 \text{ m}} = 2.1 \text{ N}$$

By definition, this force must be perpendicular to the radius, so it is the tangential force that produces the torque.

EXAMPLE—Experiment with a Rotating Platform

In a laboratory setup pictured below, an object is attached to a string that runs over a pulley and is wound around the axle of a rotating disk. When the object is released and falls, it causes the disk to rotate. (Assume the mass of the string and the mass of the pulley are small enough to be negligible.) During the experiment, multiple trials yield the following data. From these, determine the rotational inertia of the disk.

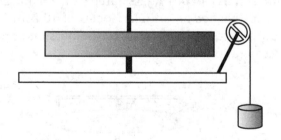

Data

Average linear acceleration of the falling object: 0.12 m/s^2

Mass of the falling object: 500 g

Radius of the axle of the rotating disk: 5.0 mm

Calculations:

Step 1. Use the acceleration and mass of the falling object to find the tension in the string. (Start with a free-body diagram of the falling object, with the direction of motion of the object taken as the positive direction.)

$$\Sigma \mathbf{F}_y = m\mathbf{a}_y$$

$$mg - T = m\mathbf{a}$$

$$T = mg - m\mathbf{a} = (0.5 \text{ kg})(9.8 \text{ m/s}^2) - (0.5 \text{ kg})(0.12 \text{ m/s}^2) = 4.84 \text{ N}$$

Step 2. Use the tension in the string and radius of the axle to find the torque exerted on the disk:

$$\tau = R_\perp \mathbf{F} = (0.005 \text{ m})(4.84 \text{ N}) = 0.0242 \text{ N} \times \text{m}$$

Step 3. Use the acceleration of the object (which is the same as the acceleration of the string) and the radius of the axle to determine the angular acceleration of the disk:

$$\mathbf{a} = \alpha R$$

$$\alpha = \frac{0.12 \text{ m/s}^2}{0.005 \text{ m}} = 24 \text{ rad/s}^2$$

Step 4. Use the torque and acceleration calculated above to determine the rotational inertia of the disk:

$$\tau = I\alpha$$

$$I = \frac{\tau}{\alpha} = \frac{0.0242 \text{ N} \times \text{m}}{24 \text{ rad/s}^2} = 0.0010 \text{ kg} \times \text{m}^2$$

EXAMPLE—Equilibrium of a Meter Stick

The uniform meter stick below has a mass of 800 g hanging at the 15 cm mark and a mass of 350 g at the 70 cm mark. It balances horizontally on a pivot placed at the 35 cm mark. What is the mass of the meter stick?

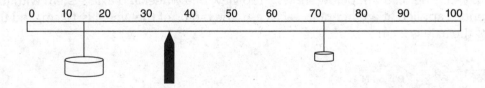

Solution:

Because this is a uniform meter stick, the mass of the stick is considered to be at its center of mass, which would be at the 50-cm mark. The pivot is already set at the 35-cm mark.

There are two clockwise torques: the weight of the stick at a distance 15 cm from the pivot and the weight of the 350-g mass at a distance 35 cm from the pivot. (Because the stick is horizontal, the forces are all perpendicular to the distances measured on the meter stick to calculate the torque.)

There is one counterclockwise torque: the weight of the 800-g mass at a distance of 20 cm from the pivot:

$$\Sigma\tau = 0$$

$$\tau_{cw} = \tau_{ccw}$$

$(0.35 \text{ kg})(9.8 \text{ m/s}^2)(0.35 \text{ m}) + (m_{stick})(9.8 \text{ m/s}^2)(0.15 \text{ m}) = (0.80 \text{ kg})(9.8 \text{ m/s}^2)(0.2 \text{ m})$

$m = 0.25 \text{ kg}$

EXAMPLE—Torque

Four equal forces (F) are labeled in the diagram below as they exert torques on a wheel and axle. The forces are all tangential to either the wheel or the axle. The wheel has a radius of 0.5 m, and the axle has a radius of 0.1 m. Rank the size of the torque exerted by each force, in order from highest to lowest, and determine the net torque in terms of F.

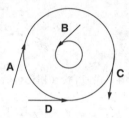

Solution:

$\tau = R_{\perp}F$ The torque from force A is 0.5 F clockwise, the torque from force B is 0.1 F counterclockwise, the torque from force C is 0.5 F clockwise, and the torque from force D is 0.5 F counterclockwise. The magnitudes of the torques is ranked: A = B = C > D.

By convention, make counterclockwise torques positive and counterclockwise torques negative. So the net torque is: 0.5 F – 0.1 F – 0.5 F + 0.5 F = 0.4 F (positive, so it is counterclockwise).

EXERCISE

8.2

Questions 1–3. The diagram below shows a top view of a wheel and axle system with forces A, B, C, and D applied at a tangent in each case. The radius of the wheel is 10 cm, and the radius of the axle is 2 cm.

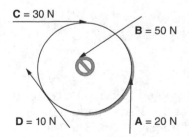

1. Which force produces the largest torque?

   A, B, C, or D

2. What is the net torque on the system?

   A. 100 N-m clockwise
   B. 100 N-m counterclockwise
   C. 200 N-m clockwise
   D. 300 N-m counterclockwise

3. Which of the following is not a vector quantity?

   A. Angular momentum
   B. Rotational inertia
   C. Angular velocity
   D. Torque

4. A thin rod 1 m in length has a 200 g object attached to one end and a 300 g object attached to the other end. At what point should the rod be supported so that it remains in horizontal equilibrium?

   A. 20 cm from the 300-g object
   B. 40 cm from the 300-g object
   C. 60 cm from the 300-g object
   D. 40 cm from the 200-g object

5. A force of 100 N is applied to a wheel as shown below. The radius of the wheel is 20 cm, and the radius of the axle is 5 cm. For the wheel to rotate at constant speed, determine the torque produced by the friction force applied to the axle that would keep it rotating at constant speed.

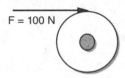

F = 100 N

   A 1.25 N-m clockwise
   B. 1.25 N-m counterclockwise
   C. 20 N-m clockwise
   D. 20 N-m counterclockwise

6. If the distance of a bicycle pedal from the axle is about 18 cm, what is the torque a bicycle rider with a mass of 50 kg would apply at the point the pedal is farthest forward and horizontal and the rider exerts all of his or her weight on the pedal?

   A. 50 N-m
   B. 88 N-m
   C. 500 N-m
   D. 1,000 N-m

7. What is the torque exerted at the elbow on a 5-kg object that is lifted from horizontal by bending at the elbow? Assume the distance from the elbow to the clenched fist is 30 cm.

   A. 10 N-m
   B. 15 N-m
   C. 100 N-m
   D. 150 N-m

8. In the situation below, the rod is 1.0 m long and has a mass of 200 g. The pivot is 20 cm to the left of the center of the rod. The object on the left end has a mass of 1.0 kg, and the object on the right end has a mass of 500 g. What is the net torque on the rod?

    A. 0.9 N-m clockwise
    B. 3.5 N-m clockwise
    C. 3.5 N-m counterclockwise
    D. 20 N-m counterclockwise

9. For an object or system to be in equilibrium:

    A. The object or system must be stationary.
    B. The net force on the object or system must be equal to zero.
    C. The net torque on the object or system must be equal to zero.
    D. Both answers B and C must be true.

10. Which of the following arrangements of three objects suspended from the horizontal rod with the pivot as shown would result in a balanced system, if Mass 1 has a mass of 1 kg, Mass 2 has a mass of 2 kg, and Mass 3 has a mass of 3 kg? The distances given are measured from the point of attachment of the object to the rod and the fulcrum.

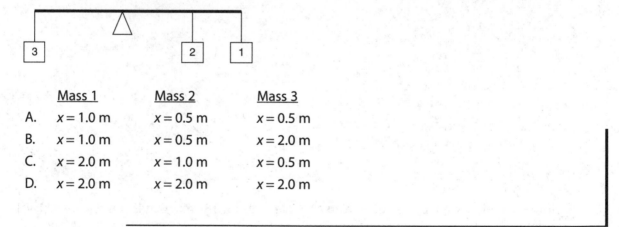

	Mass 1	Mass 2	Mass 3
A.	$x = 1.0$ m	$x = 0.5$ m	$x = 0.5$ m
B.	$x = 1.0$ m	$x = 0.5$ m	$x = 2.0$ m
C.	$x = 2.0$ m	$x = 1.0$ m	$x = 0.5$ m
D.	$x = 2.0$ m	$x = 2.0$ m	$x = 2.0$ m

Conservation of Angular Momentum

Angular Momentum, L: The product of rotational inertia and angular velocity. Angular momentum remains constant unless a net external torque is exerted on the rotating object or system:

$$L = I\omega$$

Law of Conservation of Angular Momentum: In the absence of an external torque, angular momentum of a system before and after a change within the system will remain the same.

$$L_o = L_f$$

$$I_o\omega_o = I_f\omega_f$$

EXAMPLE—Conservation of Angular Momentum

The playground merry-go-round shown below has a moment of inertia of 100 kg-m². Two children stand on the outside edge and push until they have the merry-go-round turning at 0.5 rotations per second. At that instant, the children move toward the center. What is the rotational speed when they move to the center? (Use 172 kg-m² for rotational inertia when the children are on the edge of the apparatus and 110 kg-m² for rotational inertia when the children are near the center of the apparatus.)

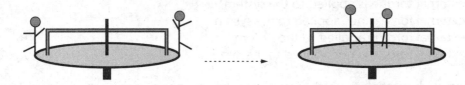

Solution:

Assuming that no external torque is exerted on the merry-go-round as the children move to the center, angular momentum is conserved. Use conservation of angular momentum to determine the new rotational speed.

$$\vec{L}_i = \vec{L}_f$$

$$I_i \omega_i = I_f \omega_f$$

(172 kg×m²)(0.5 rotation/s) = (110 kg×m²)(ω_f)

ω_f = 0.78 rotation/s

EXERCISE 8.3

1. Which of the following is a true statement about rotational inertia?

   A. The rotational inertia of an object can be increased by decreasing the object's mass but keeping the mass distribution (r) the same.
   B. The rotational inertia of an object can be increased by keeping the object's mass the same but decreasing the mass distribution (r).
   C. An object only has rotational inertia if a torque has been applied to it so that it is rotating.
   D. An object's rotational inertia can be changed by shifting position of the axis of rotation.

2. A diver jumping from a diving board curls into a "tuck" position during the dive. As a result, during the tuck, the diver's:

   A. Angular velocity increases because rotational inertia decreases.
   B. Angular velocity remains the same and rotational inertia stays the same.
   C. Rotational inertia increases and angular velocity increases.
   D. Rotational inertia remains the same but angular velocity increases.

3. A student sits on a rotating stool with arms extended outward and another student exerts a torque to start them spinning. As the student spins, they quickly bring their arms and legs inward. What is the noticeable result?

   A. Angular velocity increases because rotational inertia increases.
   B. Angular velocity remains the same because angular momentum increases.
   C. Angular velocity increases because rotational inertia decreases.
   D. Angular velocity remains the same because angular momentum decreases.

4. Angular momentum of a rotating system is conserved if:

   A. An external torque is applied to the system
   B. An external force is not applied to the system
   C. An external force is applied to the system
   D. An external torque is not applied to the system

5. An object with rotational inertia of 0.4 mr^2 is spinning at an angular velocity of 16 radians per second. The effective radius of the object is reduced by ½. What will happen to the angular velocity of the object?

   A. Angular velocity will decrease to 4 radians per second.
   B. Angular velocity will remain the same.
   C. Angular velocity will increase to 64 radians per second.
   D. Angular velocity will increase to 8 radians per second.

6. A stick has a rotational inertia of ¹⁄₁₂ mr^2 if you hold it at its center and twirl it around. Suppose you now grab the stick at its end and twirl it around its end. What would you predict about the rotational inertia of the stick?

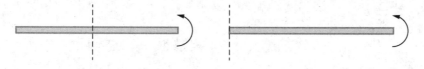

   A. It would stay the same.
   B. It would be larger.
   C. It would be smaller.

Fluids

- Density
- Pressure
- Buoyancy
- Fluid Flow and Continuity Principle
- Fluid Flow and Bernoulli's Equation

The study of fluids is not always included in an introductory course in physics, but an understanding of fluids and how they behave not only explains the real world but provides a basis for many areas of future study—particularly the health professions. In this chapter, the approach to the understanding of fluids is to use topics with which you are already familiar as examples.

Fluids: States of matter that flow, including liquids and gases

Compressible Fluids: Materials that change volume under pressure. Gases are compressible, so they don't maintain constant volume and density under pressure.

Incompressible Fluids: Most liquids, such as water, do not change volume or density appreciably under moderate pressure changes.

Viscosity: The resistance of a fluid to flow. It generally describes fluids that would be described as thick, such as honey or molasses. Viscosity in fluids is due to internal friction between layers of a moving fluid, which usually increases as temperature is lowered.

Density

Density: Mass per unit volume.

$$\rho = \frac{m}{V}$$

Specific Gravity: The ratio of density of an object or material to a standard, such as water. It is just a number with no units. For example, ice has a density of 0.9 g/ml and water has a density of 1 g/ml, so the specific gravity of ice is 0.9. Ice is less dense than water and will float in water with 90% under water and 10% above water level.

> The density of water is 1 g/ml, or 1 g/cm^3, or 1,000 kg/m^3. The density of air is about 1.29 kg/m^3 at standard conditions of temperature and pressure.

1. A liquid has a specific gravity of 0.357. What is its density?

   A. 357 kg/m³
   B. 800 kg/m³
   C. 0.357 kg/m³
   D. 643 kg/m³

2. A geologist finds that a rock sample has a mass of 12.0 g in air and displaces 6 ml of water when it is put into a container of water and sinks. What is the density of the rock?

   A. 2.5 g/ml
   B. 2.0 g/ml
   C. 1.5 g/ml
   D. 0.67 g/ml

3. A block of wood floats 60% submerged when placed in water (and thus 40% above water). When the block is placed into an alcohol that is 90% as dense as water, the block:

   A. floats 90% under the surface of the alcohol
   B. floats 67% under the surface of the alcohol
   C. floats 54% under the surface of the alcohol
   D. sinks in the alcohol

4. A material that has a density of 2.54 g/ml has what density in kilograms per cubic meter (which is most commonly used in physics)?

   A. 25.4
   B. 0.254
   C. 254
   D. 2,540

5. A small cube has a mass of 10 g and measures 2 cm on a side. What is the density of the material of which the cube is made?

   A. 1.25 g/cm³
   B. 2.5 g/cm³
   C. 5.0 g/cm³
   D. 0.80 g/cm³

Pressure

Pascal's Principle: States that the pressure in a fluid at any depth in a fluid is equal in all directions and is transmitted equally throughout the fluid at that depth.

Pressure: Force per unit area $P = F/A$; measured in N/m² or pascals (Pa).

Fluid pressure: At a given depth in a fluid is dependent upon density of the fluid (ρ), gravitational acceleration (g), and depth of fluid (h):

$$P = \rho g h$$

Fluid pressure is also measured in N/m² or pascals (Pa).

Standard Sea Level Pressure: Due to the atmosphere at Earth's surface; is 1.01×10^5 Pa.

Pascal, Pa: Unit of pressure, equal to one newton per square meter: $1\ Pa = 1\ N/m^2$.

Gauge Pressure: Pressure due to the fluid at a given depth.

$$P_G = \rho g h$$

Absolute Pressure: Total pressure at a given depth due to pressure of all fluids above a certain depth. This is usually gauge pressure added to atmospheric pressure.

$$P_{absolute} = P_{atm} + P_{gauge}$$

> **Helpful estimate:** Each 10 meter depth of fresh water is approximately equal to 1 atmosphere of pressure.

EXAMPLE—Fluid Pressure

Water is poured into a U-shaped tube. Then oil is poured into one end and the liquid levels are allowed to come to equilibrium. From the information given below, determine the density of the oil.

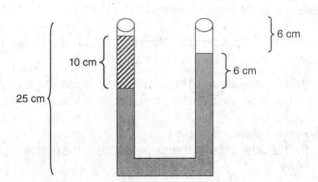

Solution:

The total fluid pressure in each side of the tube will be the same at equilibrium. The total pressure of water, plus oil, plus atmospheric pressure on the left side must equal the total pressure of water plus atmospheric pressure on the right side. Since atmospheric pressure is essentially the same on both sides, we will consider only the pressure due to oil and water:

$$P_{oil(left)} + P_{water(left)} = P_{water(right)}$$

$$\rho_{oil}\, \cancel{g}\, h_{oil} + \rho_{water}\, \cancel{g}\, h_{water(left)} = \rho_{water}\, \cancel{g}\, h_{water(right)}$$

$$\rho_{oil}\ (0.10\ m) + (1{,}000\ kg/m^3)(0.13\ m) = (1{,}000\ kg/m^3)(0.19\ m)$$

$$\rho_{oil} = 600\ kg/m^3$$

1. This tank is 8.0 m tall and has a bottom area of 50 m². The tank is filled to a depth of 6.0 m with fresh water. Determine the closest value to the absolute pressure at the bottom of the tank.

   A. 1 atm
   B. 1.5 atm
   C. 1.6 atm
   D. 2 atm

2. Compare the total pressure at the bottom of a fresh water lake that has a surface area of 100 m² and a depth of 20 m to the total pressure at the bottom of a fresh-water lake that has a surface area of 1,000 m² and a depth of 20 m.

   A. The larger lake has more total pressure at the bottom.
   B. The larger lake has less total pressure at the bottom.
   C. Both lakes have the same total pressure at the bottom.

3. A large aquarium is 6 m deep. Compare the pressure on the bottom of the aquarium to the pressure on the side walls of the aquarium near the bottom.

   A. The pressure on the side walls at the same depth is the same as pressure on the bottom.
   B. The pressure on the side walls at the same depth is less than the pressure on the bottom.
   C. The pressure on the side walls is zero.
   D. The pressure on the side walls at the same depth is more than the pressure on the bottom.

4. Determine the absolute pressure at the bottom of a fresh water lake that has a depth of 15 m and a surface area of 21,000 m². (The density of fresh water is 1,000 kg/m³, and atmospheric pressure is 101 kPa.)

   A. 101 kPa
   B. 202 kPa
   C. 150 kPa
   D. 250 kPa

5. Calculate the force (in newtons) on the top of a 1-m² section of a sunken ship at a depth of 4,000 m. (Assume the average density of seawater is 1025 kg/m³.)

   A. 1×10^5 N
   B. 4×10^6 N
   C. 4×10^7 N
   D. 1×10^8 N

6. The total fluid pressure on a scuba diver at the bottom of a lake does <u>not</u> depend upon:

    A. atmospheric pressure
    B. density of the water
    C. water depth
    D. surface area of the lake

Buoyancy

Buoyant Force: An upward force ($\mathbf{F}_B$) on an object that is partially or fully submerged in a fluid, created by the fluid due to pressure differences on upper and lower surfaces of the object. The buoyant force on an object in a fluid is proportional to the density of the fluid, the volume of fluid displaced, and gravitational acceleration.

$$\mathbf{F}_B = \rho V g$$

The buoyant force on an object immersed in a fluid is equal to the *weight* of fluid displaced by the object, so the object will only sink in the fluid as far as needed to displace its own weight. Therefore, the buoyant force upward from the fluid is equal to the gravitational force downward on an object that floats. If the objects sinks, it cannot displace enough fluid equal to its weight, but it does still have a buoyant force upward that makes the object seem to weigh less. It now has an **apparent weight** equal to the actual weight minus the buoyant force.

    Free-body diagrams are very helpful in analyzing forces on objects in a fluid. In the example on the left below, the object is partially submerged (and floating) in the fluid, and in the diagram on the right below, the object has sunk to the bottom of the container. In the first case, the buoyant force is *equal* to the weight of the object. In the second case, the buoyant force is *less than* the weight of the object. However, in both cases, the buoyant force is equal to the weight of fluid displaced by the object.

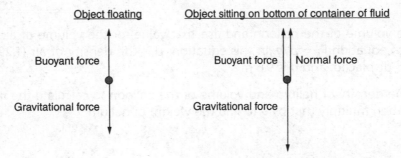

Archimedes Principle: An object immersed partially or fully in a fluid will experience an upward buoyant force from the fluid that is equal to the weight of the fluid displaced by the object.

EXAMPLE—Buoyant Force on a Balloon

In a class activity, students are given a helium-filled balloon, an empty balloon, string, meter stick, electronic balance, and a box of paper clips. They are to determine how many paper clips to attach to the balloon so that it rises as slowly as possible. Describe what measurement the students need in order to win this contest.

Solution:

Start with a free-body diagram of the balloon–helium–paper clip system.

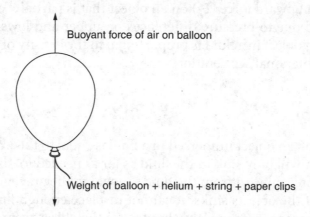

Buoyant force of air on balloon

Weight of balloon + helium + string + paper clips

For the balloon to rise most slowly, the net force on the system must be close to zero—with the buoyant force slightly larger than the total weight.

Step 1. Calculate the buoyant force. To do this, we'll need to volume of the inflated balloon. Measure the diameter in several directions and calculate the average to find an estimation of the diameter (assuming the balloon is nearly spherical). Use the radius of the balloon, in meters:

$$V = \tfrac{4}{3}\pi r^3$$

Calculate the volume of the balloon and use that value for the volume of air displaced in the buoyant force equation $F_B = \rho V g$. In this equation, ρ is the density of air (1.29 kg/m³), V is the volume of air displaced, and g is 9.8 m/s².

Step 2. Use the density of helium and volume of the balloon to calculate the mass of helium in the balloon; then multiply that by g to find the weight of helium:

$$\rho_{He} = \frac{m}{V}$$

$$\rho_{He} = (0.1786 \; \text{g}/\text{L})\left(\frac{1\,\text{kg}}{1{,}000\;\text{g}}\right)\left(\frac{1{,}000\;\text{L}}{1\,\text{m}^3}\right) = 0.1786 \; \text{kg/m}^3$$

$$m = (0.1786 \; \text{kg/m}^3)(V)$$

$$W_{He} = mg$$

Step 3. Find the total weight of all parts of the system—including the weight of helium in the balloon. Use the electronic balance to measure a length of string equivalent to that used to tie the inflated balloon and an empty balloon. Add the weight of the helium and enough paper clips to be equal to (or slightly less than) the buoyant force calculated in Step 1.

EXAMPLE—Density

A ball with a radius of 2.0 cm and a mass of 20 g floats in water. (a) What is the density of the ball? (b) What is the buoyant force on the ball? (c) How far under the water's surface will the ball float?

Solution:

(a) Determine the density of the ball using the formula and adjusting units:

$$\rho = \frac{m}{V} = \frac{m}{\frac{4}{3}\pi r^3} = \frac{0.020 \text{ kg}}{\frac{4}{3}\pi(0.02 \text{ m})^3} = 596 \text{ kg/m}^3$$

(b) If the ball floats, it is in equilibrium, with the buoyant force equal to the weight of the ball.

$$\mathbf{F}_B = mg = (0.020 \text{ kg})(9.8 \text{ m/s}^2) = 0.196 \text{ N}$$

(c) With a density of 596 kg/m³, the ball is only 59.6% as dense as water. (Water's density is 1,000 kg/m³.) The ball only has to sink far enough to displace 59.6% of its volume to displace enough water to create a buoyant force equal to its own weight.

EXERCISE 9.3

1. How is it possible for a balloon filled with helium to float in air?

    A. The buoyant force of the air on the balloon is equal to the weight of the balloon and its contents.
    B. The helium in the balloon is less dense than air.
    C. The buoyant force of the helium in the balloon creates an upward force equal to the weight of the balloon.
    D. The helium in the balloon naturally rises in air, and the weight of the balloon itself is insignificant.

2. An object that weighs 50 N in air has an apparent weight of 35 N when it is submerged in water. It doesn't displace all of its weight, so it doesn't float. The same object is then submerged in salt water. Will its apparent weight change?

    A. Yes. The object will displace the same volume of salt water, which weighs less, so the buoyant force will be less and the object will appear to lose less weight.
    B. Yes. The object will displace the same volume of salt water, which weighs more, so the buoyant force will be greater and the object will appear to lose more weight.
    C. No. The volume of the object is the same in both cases, so the same volume of water is displaced.
    D. No. The salt water is more dense, but the object displaces less of it.

3. An object is placed into a container of water and sinks to the bottom of the container. What are the forces on the object when it is at the bottom?

   A. Buoyant force upward and gravitational force downward
   B. Normal force upward and gravitational force downward
   C. Buoyant force upward, gravitational force downward, and normal force upward
   D. Normal force downward, gravitational force downward, and buoyant force upward

4. An object with mass of 100 g in air displaces 50 mL of water when placed in water. Which is true about what happens?

   A. The object will float.
   B. The object will sink and have an apparent weight of about 9 N.
   C. The object will sink and have an apparent weight of about 8 N.
   D. The object will sink and have an apparent weight of about 0.5 N.

5. An object weighs 100 g in air. How much water will it need to displace in order to float?

   A. 10 mL
   B. 50 mL
   C. 100 mL
   D. 1,000 mL

6. A fluid-filled tank is lowered into seawater. The mass of the tank and fluid is 45 kg, and it displaces 40 L of water. Determine the buoyant force on the tank.

   A. 17,640 N
   B. 440 N
   C. 17.6 N
   D. 400 N

7. A container is filled with water until it is just ready to overflow. A ball is carefully placed in the water so that the ball floats. As a result, 10 mL of water overflows. What is the weight of the ball in air?

   A. 0.98 N
   B. 0.1 N
   C. 0.01 N
   D. The weight of the ball cannot be determined unless we know what percentage of the ball floats under water.

8. What factors affect the buoyant force on an object fully or partially immersed in a fluid?

   A. Density of the object, volume of the object, and gravitational field (g)
   B. Density of the fluid, volume of fluid displaced, and gravitational field (g)
   C. Density of the fluid, volume of the object, and gravitational field (g)
   D. Weight of the object in air, volume of the object, and gravitational field (g)

9. A balloon filled with helium gas will float to the ceiling of a room on Earth. What will happen to the same balloon when released on the surface of the Moon?

   A. The gravitational force on the Moon is less, so the balloon has less weight and will rise even faster.
   B. Since there is no atmosphere, the balloon will expand to a larger volume and float easier than on Earth.
   C. Since there is no air on the Moon to provide a buoyant force on the balloon, it will sink to the surface.
   D. Since the surface is so cold, the balloon will shrink to a smaller size and thus have a lower buoyant force.

10. A solid metal cylinder attached to a string is lowered under water so it is completely submerged but is not touching the bottom. Which of the following most nearly shows the forces exerted on the cylinder?

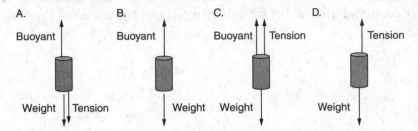

11. In the situation shown, a bar of metal is suspended under water by two cords. Which is a true statement regarding the forces on the metal bar, if T is the tension in each cord, W is the weight of the bar in air, and F is the buoyant force of the water on the bar?

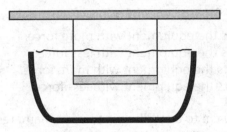

A. $F = 2T - W$
B. $2T + F = W$
C. $W > 2T - F$
D. $2T = W$

12. A ball floats halfway under the surface in a liquid. Which statement is true?

A. The ball's density is the same as the liquid's density.
B. The buoyant force on the ball is greater than the weight of the ball.
C. The buoyant force on the ball is less than the weight of the ball.
D. The buoyant force on the ball is equal to the weight of the ball.

Fluid Flow and Continuity Principle

Equation of Continuity: Defines **conservation of mass** for a moving fluid.

$$\rho_1 A_1 v_1 = \rho_2 A_2 v_2$$

In situations where the fluid does not change density, the density term cancels from both sides and the equation takes the following form, where A is cross-sectional area and v is velocity.

$$A_1 v_1 = A_2 v_2$$

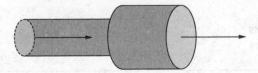

As the fluid moving from the left in this diagram enters a tube of larger diameter, the speed of fluid flow decreases. However, the volume rate of flow must stay the same.

Volume Rate of Flow: The rate of fluid flow through an opening, Q (in cubic meters per second), is dependent on the cross-sectional area, A (in square meters), and the speed of the fluid, v (in meters per second).

$$Q = \frac{V}{t} = Av$$

EXERCISE 9.4

1. Science laboratories often have spouts attached to the faucets that taper to a small opening. Why is this helpful for cleaning of equipment?

   A. The narrow spout increases water speed, so it hits the equipment with more force.
   B. The narrow spout increases water speed, so it hits the equipment with less force.
   C. The narrow spout decreases water speed, so it hits the equipment with more force.
   D. The narrow spout decreases water speed, so it hits the equipment with less force.

2. Water flows from a larger pipe with a diameter of 20 cm to a smaller pipe with a diameter of 5 cm. If the speed of the water is 4.0 m/s in the larger pipe, determine the speed of water in the smaller pipe.

   A. 8 m/s
   B. 16 m/s
   C. 64 m/s
   D. Can't determine without knowing the length of the pipe

3. The equation of continuity $A_1 v_1 = A_2 v_2$ is a statement of:

   A. Conservation of energy
   B. Conservation of velocity
   C. Conservation of momentum
   D. Conservation of mass

4. If you put your thumb over the end of a garden hose of running water, (a) what effect do you have on the speed of water flow, and (b) what effect do you have on the amount of water leaving the hose each second?

   A. Water speed and amount of water stay the same.
   B. Water speed decreases and amount of water decreases.
   C. Water speed increases and amount of water increases.
   D. Water speed increases but amount of water stays the same.

5. A trough with semicircular cross section is level full with water flowing at a speed of 3.0 m/s. If the depth of the water at the center of the trough is 0.20 m, what is the approximate volume of water flowing past a given point per hour?

   A. 120 cubic meters
   B. 680 cubic meters
   C. 1,400 cubic meters
   D. 2,200 cubic meters

Fluid Flow and Bernoulli's Equation

Bernoulli's principle: A conservation of energy statement that is basically the work-energy theorem applied to fluids.

This equation expresses the work-energy theorem for fluids. The total energy on left side of the equation, representing quantities at one point in the moving fluid, is equal to the total on the right side, representing those same quantities at a different point in the fluid.

$$P_1 + \rho g h_1 + \tfrac{1}{2}\rho v_1^2 = P_2 + \rho g h_2 + \tfrac{1}{2}\rho v_2^2$$

EXAMPLE—Bernoulli's Principle

Reexamine this diagram from the previous section where the continuity principle was applied to demonstrate conservation of mass. In this section, we will examine the same scenario using Bernoulli's equation as an example of conservation of energy. Compare qualitatively the kinetic energy of the fluid in the two sections of the pipe and the pressure of the fluid on both sections, using Bernoulli's principle.

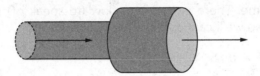

Solution:

$$P_1 + \rho g h_1 + \tfrac{1}{2}\rho v_1^2 = P_2 + \rho g h_2 + \tfrac{1}{2}\rho v_2^2$$

For this situation, the left side of the equation will describe total energy in the narrow pipe, and the right side will describe total energy in the wider portion of the pipe.

The $\rho g h$ terms on both sides will cancel, since the depth of fluid is not significantly different. We're left with:

$$P_1 + \tfrac{1}{2}\rho v_1^2 = P_2 + \tfrac{1}{2}\rho v_2^2$$

We have already determined from the continuity principle in the previous section that the speed of fluid flow is less in the wider pipe and greater in the narrower pipe. Therefore, the ½ ρv^2 term is greater in the narrow pipe and less in the wider pipe.

This means for the equation to be true, the fluid pressure in the narrower pipe must be less than in the wider pipe. We come to this conclusion:

When the speed of fluid flow increases, the fluid pressure decreases.

EXAMPLE—Bernoulli's Principle

An open container of water has a spout 50 cm from the bottom that allows a stream of water to flow out of the container. (a) When the height of water above the spout is 40 cm, determine the speed of water flowing out of the spout. (b) Calculate how far from the base of the container the water will land. (c) How does the distance from the container that water lands change as the water level is lowered?

Solution:

Using Bernoulli's Equation, we can assume the air pressure inside and outside of the container is approximately the same. There is no water flow (to speak of) inside the container and no water pressure outside, so we'll cancel those terms, leaving:

(a) $P_i + \rho g h_i + \frac{1}{2}\rho v_i^2 = P_o + \rho g h_o + \frac{1}{2}\rho v_o^2$

$\rho g h_i = \frac{1}{2}\rho v_o^2$

$(9.8 \text{ m/s}^2)(0.40 \text{ m}) = \frac{1}{2}v^2$

$v = 2.8 \text{ m/s}$

Now that we know the horizontal speed as the water leaves the spout, we can just treat this as a projectile motion problem. Basically, each droplet of water is just an object in projectile motion as it leaves the spout.

(b) Vertical motion: $\Delta y = v_{yi}t + \frac{1}{2}gt^2$

$-0.5 \text{ m} = 0 + \frac{1}{2}(9.8 \text{ m/s}^2)t^2$

$t = 0.32 \text{ s}$

Horizontal motion: $\Delta x = v_x t$

$\Delta x = (2.8 \text{ m/s})(0.32 \text{ s}) = 0.90 \text{ m}$

(c) As the water height decreases, water pressure also decreases, so the velocity of the water coming out of the spout will also decrease and land less far from the container.

EXAMPLE—Bernoulli's Principle and an Airfoil

The speed of air moving across the top of an airfoil is 12 m/s, and the speed under the airfoil is 9 m/s. If the area perpendicular to the air flow is 10 m², calculate the lift force due to the pressure difference.

Solution:

Use Bernoulli's principle to calculate the difference in pressure between upper and lower air layers. The assumption we can make in this case is that the airfoil is thin, so the difference in pressure due to depth of air (ρgh) is negligible, so we will cancel those terms.

$$P_A + \rho g\cancel{h_A} + \tfrac{1}{2}\rho v_A^2 = P_B + \rho g\cancel{h_B} + \tfrac{1}{2}\rho v_B^2$$

$$P_A + \tfrac{1}{2}(1.29 \text{ kg/m}^3)(12 \text{ m/s})^2 = P_B + \tfrac{1}{2}(1.29 \text{ kg/m}^3)(9 \text{ m/s})^2$$

It's easy to see that if the velocity is higher above the wing—by conservation of energy—the pressure will be higher below the wing. We don't need to know what the pressures are above and below, just the pressure difference. From this, we can calculate the force:

$$\Delta P = P_B - P_A = 40.6 \text{ Pa}$$

$$\Delta P = \frac{\mathbf{F}}{A}$$

$$\mathbf{F} = \Delta PA = (40.6 \text{ Pa})(10 \text{ m}^2) = 406 \text{ N}$$

EXERCISE 9.5

1. Bernoulli's Equation is a statement of:

   A. conservation of energy
   B. conservation of velocity
   C. conservation of momentum
   D. conservation of mass

2. When air flows over the top of an airplane wing, it moves faster than the air moving under the wing. According to Bernoulli's equation, this means that:

   A. The pressure in the moving air above the wing is less than the pressure below the wing, resulting in lift.
   B. The pressure in the moving air below the wing is less than the pressure above the wing, resulting in lift.
   C. There is essentially no pressure difference above and below the wing due to moving air.

3. A fluid flows from a wider tube into a thinner tube. What happens to the velocity and pressure of the fluid in the thinner tube?

    A. velocity increases and pressure increases.
    B. velocity increases and pressure decreases.
    C. velocity decreases and pressure increases.
    D. velocity decreases and pressure decreases.

4. When you put your thumb over the end of a garden hose on the ground and point it upward, the water will project up into the air. What is the best explanation?

    A. Water pressure pushes water molecules out of the hose in every direction.
    B. The fast moving water coming out of the hose has more pressure to push it upward.
    C. The kinetic energy of the water coming from the hose is converted to gravitational potential energy.
    D. The potential energy of the water inside the hose is converted to kinetic energy.

5. Which of the following expressions could be used to calculate how fast water will run out of the bottom of a tank if we know the height of the water in the tank?

    A. $\rho gh = \frac{1}{2} \rho v^2$
    B. $\frac{1}{2} \rho gh = \rho v^2$
    C. $mgh = \frac{1}{2} \rho v^2$
    D. $\rho gh^2 = \frac{1}{2} \rho v^2$

Thermodynamics

- **Kinetic Theory and Temperature**
- **Ideal Gas Law**
- **Heat, Work, and Conservation of Energy**
- **Thermodynamic Systems**

Thermodynamics is really a small part of an entire unit in temperature, kinetic theory of molecules and the laws of thermodynamics. The fundamentals of kinetic theory are provided here, along with an overview of ideal gases and conservation of energy in thermodynamic processes. However, Maxwell's distribution and entropy are left for more advanced course work, and only basic pressure versus volume diagrams are used to illustrate thermodynamic processes.

Kinetic Theory and Temperature

The **Kinetic Theory** describes the constant, random motion of molecules. Kinetic energy increases as the absolute temperature of the gas increases.

Temperature, T: A measure of the average translational kinetic energy of molecules. When used in equations related to gas laws, the absolute or Kelvin temperature (K) should be used, since all temperatures on that scale would be positive. Add 273 to Celsius temperatures to obtain temperature in kelvins.

$$KE = \frac{3}{2} k_B T$$

where k is Boltzmann's constant and T is temperature in kelvins.

Absolute Zero: The temperature at which all molecular motion ceases. This is defined as zero degrees Kelvin (0 K) or −273°C or −459°F. (Note: Degrees are used with measurements on both the Celsius).

Heat or Heating: Transfer of energy from a system with higher temperature to a system with lower temperature—not a property of the system or the amount of energy contained in the system.

Thermal Energy: The energy contained in a system due to kinetic energy of molecules (considered the same as **internal energy** in most cases).

NGSS HS-PS3-4

Specific Heat, *c*: Describes the amount of energy required to raise the temperature of a certain amount of a material by one degree. For example, the specific heat of steam is 0.5 calories per gram per Celsius degree. So it would require 100 calories to raise the temperature of 10 grams of steam from 100°C to 120°C.

$$Q = mc\Delta T$$

EXERCISE
10.1

1. The energy transfer from a system at higher temperature to a system at lower temperature is called:

   A. thermal energy
   B. kinetic energy
   C. heat
   D. internal energy

2. The energy contained in a system due to transfer of energy to it is called:

   A. thermal energy
   B. heat
   C. temperature
   D. heat capacity

3. Temperature is best described as:

   A. the total motion of molecules
   B. proportional to the total energy of molecules
   C. proportional to the kinetic energy of molecules
   D. The heat capacity of a system of molecules

4. If the absolute or Kelvin temperature of an ideal gas is doubled, what is the change in the average velocity of gas molecules?

   A. The average velocity of gas molecules will not change.
   B. The average velocity will be doubled.
   C. The average velocity will be four times as much.
   D. The average velocity of molecules will be multiplied by square root of two.

5. A container of ideal gas consists of a cylinder with a moveable piston that does not allow gas molecules to enter or leave as the piston is moved. The pressure in the container is *P* and the volume is *V* before the piston is moved. The piston is then moved slowly so that the temperature does not change as the volume is reduced to *V*/2. What is the best explanation for what happens to the gas pressure, in terms of molecular motion?

   A. The average speed of molecules is greater, so they exert more force on the walls of the container.
   B. The average speed of molecules stays the same, but they collide with the walls of the container more often, increasing pressure.
   C. The average speed of molecules is less, but they collide with the walls of the smaller container more often.
   D. The average speed of molecules does not change, so the pressure inside the container decreases.

6. A container of ideal gas consists of a cylinder of constant volume that does not allow gas molecules to enter or leave. The pressure in the container is *P* and the temperature is *T* before the cylinder is heated. What is the best explanation for what happens to the gas pressure, in terms of molecular motion?

A. The average speed of molecules increases, so they exert more force on the walls of the container upon collision (more impulse), increasing pressure.
B. The average speed of molecules stays the same, but they collide with the walls of the container more often, increasing pressure.
C. The average speed of molecules is less, but they collide with the walls of the smaller container more often, increasing pressure.
D. The average speed of molecules does not change, so the pressure inside the container decreases.

7. A steel bridge is 30.0 m long when the temperature is 0°C. If the coefficient of linear expansion for this type of steel is $7.2 \times 10^{-6}/C°$, what is the length of the bridge on a day when the temperature is 40°C?

A. 30.1 m
B. 30.009 m
C. 29. 001 m
D. 31.0 m

Ideal Gas Law

Ideal gas molecules are in constant motion, with behavior that is predictable under these assumptions:

- Molecules are point particles (i.e., no rotations or oscillations along molecular bonds)

- Molecules are spaced far apart (i.e., no intermolecular attractions or repulsions)

- All collisions are perfectly elastic (i.e., both momentum and kinetic energy are conserved)

- Large number of molecules

Pressure, *P*: Defined as force exerted on an area, measured in pascals (Pa) or newtons per square meter.

$$P = \frac{F}{A}$$

The **Ideal Gas Law** relates the **state variables** for an ideal gas, using the following most common forms:

(1) $PV = nRT$

where *P* is pressure, *V* is volume, *n* is the number of moles, *R* is the gas constant, and *T* is absolute temperature in kelvins.

(2) $PV = NkT$

where P is pressure, V is volume, N is the number of molecules, k is Boltzmann's constant, and T is absolute temperature in kelvins.

Simplifications of the Ideal Gas Law

- If temperature is held constant during a change in the gas system, **Boyle's Law** applies:

$$P_1V_1 = P_2V_2$$

- If volume does not change (such as in a sealed container) during a change in the gas system:

$$\frac{P_1}{T_1} = \frac{P_2}{T_2}$$

- If pressure is held constant during a change in the gas system, **Charles's Law** applies:

$$\frac{V_1}{T_1} = \frac{V_2}{T_2}$$

EXERCISE
10.2

1. In a sealed, rigid container of an ideal gas held at constant volume, one would expect:

   A. pressure to increase as temperature increases
   B. the average kinetic energy of molecules to decrease as pressure increases
   C. the internal energy of molecules to decrease as temperature increases
   D. pressure to decrease as temperature increases

2. A rigid container of an ideal gas is heated so that the absolute temperature of the gas doubles while the volume remains constant. The molecules of the gas:

   A. increase in pressure, so they do more work on the walls of the container
   B. have twice as much internal energy
   C. must give off energy, since they cannot expand
   D. have twice the average speed as before

3. An ideal gas is one that behaves according to a set of relationships and equations, such as the ideal gas law, $PV = nRT$. Which of the following statements does <u>not</u> necessarily define an ideal gas?

   A. All collisions of molecules within an ideal gas are elastic.
   B. A system of an ideal gas consists of a large number of molecules.
   C. Ideal gas molecules are point particles.
   D. Ideal gas molecules always expand isothermally when pressure is decreased.

4. Two chambers of the same ideal gas are held at the same temperature and same constant volume, but Chamber B has twice as many gas molecules as Chamber A. Which of the following correctly describes pressure?

   A. Chamber B will have twice the gas pressure, since all molecules have the same speed but there are twice as many collisions with the walls of the container.
   B. Chamber B will have twice the gas pressure, since the molecules will have higher speed and hit the walls of the container with more force.
   C. Chamber A will have twice the gas pressure, since pressure is inversely proportional to the number of molecules.
   D. Chamber A will have twice the gas pressure, since there are fewer molecules and they can move farther before hitting the walls, they can exert more force.

5. In an ideal gas, which of the following is assumed to be true?

   A. There are few enough molecules that their motions can be easily tracked.
   B. Collisions of gas molecules with the walls of the container must be inelastic.
   C. Collisions between molecules are considered to be elastic.
   D. Molecules are large enough to exert large forces on the container walls when they change momentum.

6. An air bubble that has a volume of 0.001 m^3 is released at a depth of 21 m (2 atm) in a freshwater lake. The volume of the bubble when it reaches the surface is nearest to:

   A. 0.001 m^3
   B. 0.002 m^3
   C. 0.003 m^3
   D. 0.004 m^3

7. In a sealed, rigid container of an ideal gas held at constant volume, you would expect:

   A. pressure to increase as temperature increases
   B. average kinetic energy of molecules to decrease as pressure increases
   C. internal energy of molecules to decrease as temperature increases
   D. pressure to decrease as temperature increases

Heat, Work, and Conservation of Energy

Heat, Q: Transfer of energy from a region of higher temperature to a region of lower temperature. The rate of this transfer through a material is dependent upon thermal conductivity, k, cross-sectional area, A, thickness of material, d, and difference in temperature between the two reservoirs, ΔT.

$$\frac{Q}{t} = \frac{kA\Delta T}{d}$$

Work, W: The product of pressure and change in volume. There is some disagreement among textbook authors about how to use the equation for work. Sign convention 1 is used in this book, but both sign conventions are correct, as long as you are clear about what they mean.

Sign convention 1: The negative sign on the equation below is necessary to make work done *on* the system, which reduces volume, equal to positive work, *according to the convention used in this workbook.**

$$W = -P\Delta V \quad \text{and} \quad \Delta U = Q + W$$

Using these equations, positive work is done *on* the gas. Thus a reduction in gas volume (which is negative) means positive work was done on the gas. When a gas expands, work is done *by* the gas molecules, which is negative according to this convention. The sign convention used here is consistent with "work done on a system increases its energy."

For example, if the molecules of a system do work (perhaps expanding their container) without any heat transferred in or out ($Q = 0$), they do negative work, so W is negative and then ΔU is negative—meaning that the molecules have used their own internal energy to do work on their surroundings.

In another example, if work is done *on* a system of molecules (perhaps compressing the container), the volume decreases, but work is positive because the equation for work has a negative sign. So external forces do positive work on the system. (This is consistent with other systems, such as a spring, where doing positive work on the system—such as compressing it—increases its potential energy.)

Alternate sign convention 2: However, many textbooks use a different sign convention, where thermodynamics systems do positive work on their surroundings. It can be confusing to students. In the opposite case, the above equations will take different forms:

$$\Delta U = Q - W \quad \text{and} \quad W = P\Delta V \quad \text{only if work done on a system is defined as negative}$$

Internal energy, U: Total energy of gas molecules, which is dependent upon temperature in kelvins:

$$U = \frac{3}{2}nRT$$

In the equation, n is number of moles of gas, R is the ideal gas constant, and T is Kelvin temperature. At this level, we will consider thermal energy and internal energy to be the same.

Conduction: Transfer of energy from molecule to molecule. (Keep in mind that the energy is transferred by the molecules, though they vibrate or stay in place.) Rate of heat transfer depends on *thermal conductivity* of the material k, cross-sectional area through which the heat is conducted A, temperature difference between the chambers ΔT, and is inversely proportional to the length L, through which the heat travels:

$$\frac{Q}{t} = \frac{kA\Delta T}{L}$$

Convection: Transfer of energy by movement of fluids at different temperatures and densities. Warm fluid is less dense and rises, while cooler fluid is more dense and sinks.

Radiation: Transfer of energy by electromagnetic waves (generally considered to be in the infrared range).

Thermal Expansion: Increase in length or volume of a material due to change in temperature. The amount a material changes in length or volume with temperature depends on the properties of the material. A metal rod, for example, will increase in length an amount ΔL based upon the **coefficient of linear thermal expansion**, α, of that material, the original length of the rod, L, and the change in temperature, ΔT: $\Delta L = \alpha L_0 \Delta T$ (the units depend on the units of α).

First Law of Thermodynamics: A conservation of energy statement where change in internal energy, ΔU, is equal to the sum of work done on or by the gas (W) and heat transfer in or out of the gas (Q):

$$\Delta U = Q + W$$

Using this form of the First Law, heat transferred into the gas is considered to be positive and heat transferred out of the gas is negative. Work done on the gas by an external force is positive, and work done by the molecules of the gas is negative.

Isothermal: Thermodynamic process during which there is no change in temperature. Since change in internal energy depends upon change in temperature, ΔT and ΔU are both zero. Since $PV = nRT$ and n, R, and T are constant, pressure times volume is constant everywhere along the curve.

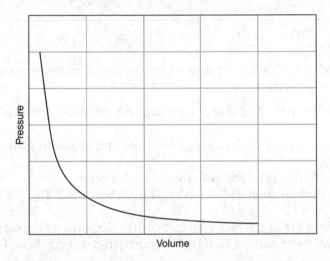

Isobaric: Thermodynamic process that takes place at constant pressure.

Constant Volume Process: Since work depends on change in volume, no work is done on or by the system during a process in which there is no change in volume.

Adiabatic: Describes a thermodynamic process during which no heat is transferred in or out of the system (i.e., $\Delta Q = 0$). On a pressure versus volume diagram, an adiabatic curve looks very much like an isothermal curve, but the adiabatic curve does not maintain constant temperature. Below is a comparison:

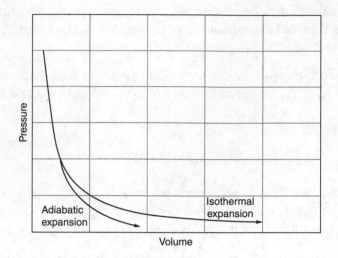

1. The rate at which heat flows by conduction from a hot chamber to a cold chamber through a steel rod may be increased by:

   A. Substituting a rod with the same dimensions but made of a material with higher specific heat
   B. Decreasing the temperature of the hot chamber and increasing the temperature of the cold chamber
   C. Substituting a shorter steel rod with the same diameter
   D. Substituting a thinner steel rod of the same length

2. Two isolated regions are in contact with each other, Region 1 at temperature T_1 and Region 2 at a lower temperature T_2. If heat is transferred directly from Region 1 to Region 2:

   A. The net amount of heat transferred out of Region 1 is greater than the heat transferred into Region 2.
   B. The net amount of heat transferred out of Region 1 is less than the heat transferred into Region 2.
   C. The net amount of heat transferred out of Region 1 is equal to the amount of heat transferred into Region 2.

3. Which of the following is a true statement for all situations regarding change in internal energy, ΔU, of a system of an ideal gas?

   A. Internal energy will increase if work is done on the system during an isobaric process.
   B. Internal energy will decrease if work is done by the system during an isovolumetric process.
   C. Internal energy will increase if heat is removed from the system and work is done on the system.
   D. Internal energy will increase if heat is added during an isothermal process.

4. An ideal gas is taken through one step of a thermodynamic cycle. During that step, which of the following would not change the average kinetic energy of the gas molecules?

   A. an isothermal decrease in volume
   B. an increase in pressure at constant volume
   C. an adiabatic expansion at constant volume
   D. an isobaric expansion

5. When an ideal gas is taken through one step of a thermodynamic cycle, which of the following terms applies if there is no heat added or removed?

   A. isothermal
   B. adiabatic
   C. isovolumetric
   D. isobaric

Thermodynamic Systems

In this section, we will examine how the equations presented earlier in this unit can be combined to analyze thermodynamic processes. They start with diagrams that plot Pressure versus Volume for a complete cycle. Then each step of the cycle, designated by arrows, can be analyzed in regard to changes in temperature, pressure, volume, and energy.

The Laws of Thermodynamics:

- Zeroth Law: Heat will flow from a system at higher temperature to a system at lower temperature until thermal equilibrium.

- First Law (conservation of energy):

  $Q = \Delta U - W^*$

- Second Law (entropy): Isolated thermodynamic systems tend to evolve toward randomness and energy tends to be unavailable to do work.

*Important note: The sign convention used here is consistent with "work done on a system increases its energy" as described in sign convention 1 in the previous section.

EXAMPLE (Advanced)—Pressure/Volume Diagram Analysis and Isobaric/Isothermal Processes

One mole of an ideal gas is taken through the process A-B-C-D as shown below. (a) Which step, if any, is isobaric? (b) Which process (if any) is isothermal? (c) Which process (if any) isovolumetric? (d) Calculate the work done during each step and for the cycle. (e) Calculate the change in temperature for each step in the cycle and for the entire cycle. (f) Calculate the heat transfer for the entire cycle.

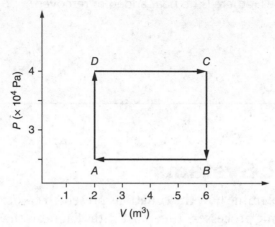

Solution:

(a) An isobaric process is one undergone at constant pressure, which would be both steps BA and DC.

(b) An isothermal process is one undergone at constant temperature. None of the steps are isothermal, since either pressure or volume changes in each step while the other does not change. By $PV = nRT$, a change in either P or V alone cannot happen without a joint change in T.

(c) An isovolumetric process is one at constant volume. Steps AD and CB occur at constant volume.

(d) Since work is the product of pressure and change in volume, the area between the line and the volume axis is the amount of work.

In step BA: $W = -P\Delta V = -(1 \times 10^4 \text{ Pa})(-0.4 \text{ m}^3) = 4 \times 10^3 \text{ J}$.

In step AD: $\Delta V = 0$, so no work is done during this step in the cycle.

In step DC: $W = -P\Delta V = -(4 \times 10^4 \text{ Pa})(0.4 \text{ m}^3) = -1.6 \times 10^4 \text{ J}$.

In step CA: $\Delta V = 0$, so no work is done during this step in the cycle. The net work done during the cycle can be determined by adding to find the total for the above four steps, which is 1.2×10^4 joules. It is also the area enclosed by the cycle.

(e) The easiest way to do this is to first use the ideal gas law:

$$P_A V_A = nRT_A$$

to find the temperature at one point. $(1 \times 10^4 \text{ Pa})(0.2 \text{ m}^3) = (1 \text{ mole})(8.31 \text{ J/mole·K})(T)$

$$T = 241 \text{ K}$$

The same gas law and constants apply to every point, so we can equate all the points:

$$\frac{P_A V_A}{T_A} = \frac{P_B V_B}{T_B} = \frac{P_C V_C}{T_C} = \frac{P_D V_D}{T_D}$$

$$\frac{(1 \times 10^4 \text{ Pa})(0.2 \text{ m}^3)}{241 \text{ K}} = nR$$

$$T_B = 723 \text{ K}$$
$$T_C = 2{,}892 \text{ K}$$
$$T_D = 964 \text{ K}$$

Because the gas is back to its original point after the complete cycle, the temperature is back to its original 241 K, and there is no net change in temperature for the cycle.

(f) We now know enough about the cycle to apply the First Law of Thermodynamics, which is a conservation of energy statement:

$$\Delta U = Q + W$$

$$\Delta T = 0, \text{ so } \Delta U = 0$$

$$W = \text{area} = 1.2 \times 10^4 \text{ J}$$

$$Q = \Delta U - W = 0 - (1.2 \times 10^4 \text{ J}) = -1.2 \times 10^4 \text{ J}$$

This negative value for Q indicates that heat is removed during the process.

EXAMPLE—PV Diagram

Four-tenths mole of an ideal gas at point **A** undergoes an increase in pressure at constant volume until it reaches the conditions at point **B**. Then the gas expands to point **C**. From point **C** back to point **A**, the gas is compressed at constant pressure. The thermodynamic process is described by the *pressure versus volume* diagram on the next page.

(a) Determine the temperature at point A.

(b) What type of process is CA?

(c) How much work is done in process AB?

(d) How much work is done in process CA? Is it done by the gas or on the gas?

(e) Estimate the net work done for the entire cycle.

(f) What is the change in internal energy for the entire cycle?

(g) Is there a net heat loss or heat gain during the entire cycle?

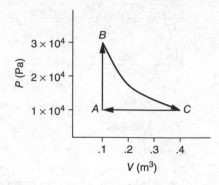

Solution:

(a) Use the ideal gas law and values from the graph to determine the temperature at A.

$PV = nRT$

$(1 \times 10^4 \text{ Pa})(0.1 \text{ m}^3) = (0.4 \text{ mole})(8.31 \text{ J/mol·K})T$

$T = 301 \text{ K}$

(b) The pressure is constant from C to A, so it is isobaric.

(c) From A to B, there is no change in volume, so no work is done in that step of the process.

(d) The work from C to A is the area under the line: $W = \text{area} = (1 \times 10^4 \text{ Pa})(0.3 \text{ m}^3) = 3{,}000 \text{ J}$. Since volume is decreased, the work is done *on* the gas, so the work is positive.

(e) The net work for the entire cycle is the area inside the loop. We'll approximate that as a triangle: $\frac{1}{2} (2 \times 10^4 \text{ Pa})(0.3 \text{ m}^3) = 3{,}000 \text{ J}$. Since there is more expansion of the gas than compression during the three steps of the cycle, the net work is done *by* the gas, so it is negative.

(f) Since the gas returns to its original pressure and volume, it must also be back to its original temperature. If there is no net change in temperature, there is no net change in internal energy.

(g) $\Delta U = Q + W$

$0 = Q + (-3{,}000 \text{ J})$

Therefore, $Q = 3{,}000 \text{ J}$. This positive answer means more heat was transferred into the gas than out of the gas during the process.

EXAMPLE—Laboratory Experiment

Students set up a laboratory experiment to determine "Joule's Heating Equivalent." In other words, they are to determine an equivalence between electrical energy input to a system (in joules) and the heat input to a material to raise its temperature (in calories). The students know that one calorie equals 4.186 joules, so this is what they set out to demonstrate.

They connect a power supply to an immersible heating element and submerge the element in a container of water. An ammeter and voltmeter are connected to read electric current and potential difference. A temperature probe is immersed in the water to read temperature change. The container is sealed as well as possible to reduce energy loss.

(a) Calculate the electrical energy input, in joules.

(b) Calculate the energy absorbed by the water, in calories.

(c) Discuss how the lab could be conducted to reduce energy loss by conduction, convection, and radiation.

Data: Potential difference (V) = 6.0 v

          Current (I) = 6.2 mA

          Specific heat of water (c) = 1 cal/g-°C

          Mass of water (m) = 200 g

          Time = 5.00 min

          Change in temperature (ΔT) = 3.5C°

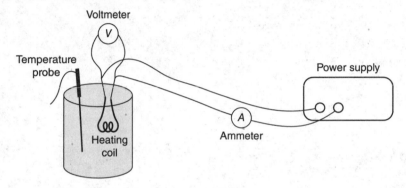

Solution:

(a) Electrical input:

$$P = VI = (6 \text{ v})(0.0062 \text{ A}) = 0.0372 \text{ w}$$

$$P = \frac{E}{t}$$

$$E = Pt = (0.0372 \text{ W})(300 \text{ s}) = 11.2 \text{ J}$$

(b) Heat energy output:

$$Q = mc\Delta T = (0.2 \text{ kg})(4.186 \text{ J/kg} \times C°)(3.0C°) = 2.51 \text{ cal}$$

Theoretically, with no loss of energy from the system, the heating output should be:

$$(11.2 \text{ J})\left(\frac{1 \text{ cal}}{4.186 \text{ J}}\right) = 2.68 \text{ J}$$

(c) Putting the container of water into a Styrofoam container could reduce loss due to conduction. Lining the inner surfaces with foil or putting ink into the water might reduce radiation losses. Sealing the top of the container of water would reduce loss of energy to air as warmed water rises to the top by convection.

Electric Charge, Field, Force, and Potential

- Electric Charge and Charge Distribution
- Electric Field
- Electric Force
- Electric Potential and Potential Energy

Electric Charge and Charge Distribution

Electric Charge: Represented by the symbols Q or q, electric charge is a property of matter. The charge on one electron (e^-), is 1.6×10^{-19} coulombs. Like charges (positive-positive or negative-negative) exert forces of repulsion on each other, and unlike charges (positive-negative) exert forces of attraction on each other.

Coulomb's Constant, k, can be used interchangeably with $1/4\pi\varepsilon_o$. Both are equal to 9×10^9 N-m^2/C^2.

Insulator: A material that does not conduct charge or electricity easily.

Conductor: A material that conducts electric charge and electric current easily.

Grounding: Connecting a charged or polarized object to the ground or a large neutral source so that excess charge can move to or from the object.

Electrostatic Equilibrium: Condition where charge has moved to all parts of a system to distribute charge evenly.

Neutral Object: An object with no net charge. All objects contain charges (protons and electrons), but **positively** charged objects have fewer electrons than protons and **negatively** charged objects have excess electrons. Only the electrons move during charging and discharging electrostatically.

Polarization: Separation of positive and negative charges on an object, though the overall net charge on the object is still zero.

Induction: In the context of electrostatics and charging, induction is the process of causing a charge separation or causing a net charge on an object by bringing a charged object near to an neutral object so that the charge separates on the neutral object. Grounding the second object while the charges are separated and then removing the original charged object will leave the second object with a net charge.

NGSS HS-PS2-6 and HS-PS3-1 and HS-PS4-5

1. How many electrons are transferred in the process of charging a latex balloon to
   1.6×10^{-8} C?

   A. 1×10^{11}
   B. 2.56×10^{-27}
   C. 1×10^{27}
   D. 1×10^{-25}

2. A metal sphere with a charge of $+2Q$ comes into contact with a metal sphere of identical
   size that has a charge of $-4Q$. Then the spheres are separated. What are the charges on the
   spheres after they are separated?

   A. Both spheres now have zero net charge.
   B. Each sphere now has a charge of $-2Q$.
   C. Each sphere now has a charge of $-Q$.
   D. Each sphere retains its original charge.

3. A negatively charged rod is brought near a second rod that is neutral and is suspended by
   a nonconducting string. The suspended rod begins to move toward the negative rod,
   showing attraction of the two rods. After the charged rod is removed:

   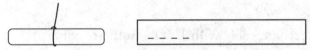

   A. The second rod has no net charge on it.
   B. The second rod now has a positive net charge.
   C. The second rod now has a negative net charge.
   D. The second rod is polarized, with one end negative and one end positive.

4. A negatively charged rod is brought near a second rod that is neutral and is suspended by
   a nonconducting string. A wire is connected from the second rod to the ground. Then,
   keeping the first charged rod in place, the ground wire is cut. The first charged rod is
   removed. After the first rod is removed:

   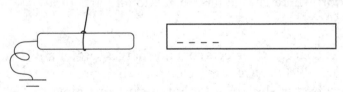

   A. The second rod has no net charge on it.
   B. The second rod now has a positive net charge.
   C. The second rod now has a negative net charge.
   D. The second rod is polarized, with one end negative and one end positive.

5. A solid metal sphere that sits on an insulating stand is given a net charge of -10 μC. Which
   statement best describes the charged sphere?

   A. The net charge will be distributed evenly throughout the volume of the metal sphere.
   B. The net charge will be distributed evenly over the surface of the sphere.
   C. The net charge will concentrate on the side of the sphere near the insulating stand.
   D. The net charge will be distributed evenly with half the charge on the outside of the
      sphere and half the charge on the inside of the sphere.

Electric Field

Electric field, *E***:** A vector quantity that defines a region in space in all directions around a charge where an electric force is exerted on other charges. Electric field around a positive charge is defined as radially outward in all directions, and electric field around a negative charge is radially inward.

$$E = \frac{1}{4\pi\varepsilon_o} \frac{Q}{r^2} = \frac{kQ}{r^2}$$

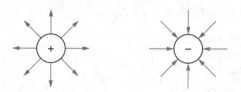

The electric field in the vicinity of a charged object is represented by vectors that diminish in size as they are drawn farther from the charge. These vectors represent a vector field that extends infinitely through space, getting weaker with distance (according to the **inverse square law**).

The electric field between metal plates that have equal and opposite charge is also represented by vectors. In this case, it is a **uniform electric field**, meaning it has the same strength and direction at every point between the plates (except some variation near the edges of the plates). This situation will come up again as you study parallel-plate capacitors in the next chapter.

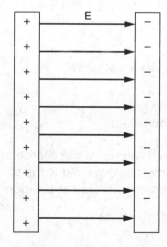

When equilibrium is reached and there is no longer movement of charges on a conducting material (such as a metal sphere or the metal shell of an airplane or a car), the electric field vectors "cancel" inside the conductor. As a result, *for any conductor at equilibrium,* ***there is no electric field inside the conductor.*** All the excess charges distribute evenly on the outside of the conductor.

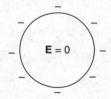

Questions 1 and 2 use this diagram.

Two charged particles are held in position as shown, with a +2.0 μC charge at $x = 1$ m and a −4.0 μC charge at $x = 6$ m.

+2.0 μC   −4.0 μC

1. Determine the net electric field at $x = 3$ m.

   A. 8,500 N/C to the right
   B. 4,500 N/C to the right
   C. 4,000 N/C to the right
   D. 500 N/C to the left

2. Where could the net electric field due to the two charges equal zero?

   A. At some point between the two charges
   B. At some point left of $x = 1$ m
   C. At some point to the right of $x = 6$ m
   D. There is no such position.

3. Two parallel metal plates are charged with $+Q$ on the top plate and $−Q$ on the bottom plate. What are the directions between the plates of the electric field, the increase in electric potential and the electric force on a proton?

   ++++++++++++++++++

   A. upward, upward, downward
   B. upward, downward, upward
   C. downward, upward, downward
   D. downward, downward, downward

4. (Advanced) Two particles, each carrying a net charge of +6 μC, are placed on the x-axis at $x = −3$ m and $x = +3$ m as shown below. What is the direction of the net electric field due to the two charges at point P, which is located at position (0, +3 m)?

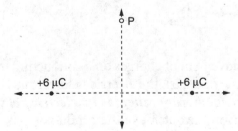

   A. to the right
   B. upward on the page
   C. downward on the page
   D. to the left

5. Determine the magnitude of the electric field at a distance of 2.0 cm from the center of an object with a charge of 2 nC.

   A. 4,500 N/C
   B. 9,000 N/C
   C. 45,000 N/C
   D. 90,000 N/C

6. The diagram shows two parallel charged metal plates. Which statement below is <u>not</u> true regarding the situation shown?

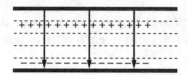

   A. The arrows represent the direction of the electric field between the plates.
   B. The horizontal dashed lines represent equipotential surfaces.
   C. The top plate is at higher electric potential than the bottom plate.
   D. The electric field increases in magnitude moving upward in the diagram.

7. The diagram below shows two charged particles held in position along a line. What is the direction of the net electrical field at point P due to the two charges?

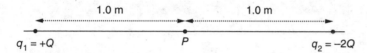

   A. to the left
   B. to the right
   C. out of the page
   D. into the page

Electric Force

Electric Force: Electric force can be described as the product of the electric field (from the source charge) and the size of the test charge. The vector equation below describes the size and direction of such a force:

$$\vec{F} = q\vec{E}$$

> Electric forces and electric fields are *vectors* and must be combined by addition of components—not by simple addition.

Coulomb's Law: The law as described by the equation below, where the force between two charged objects Q and q is directly proportional to the product of the charges and inversely proportional to the square of the distance between them.

$$F = \frac{1}{4\pi\varepsilon_o}\frac{Qq}{r^2} = \frac{kQq}{r^2}$$

Coulomb's constant: The proportionality constant, k, equal to 9×10^9 N-m^2/C^2 or $\frac{1}{4} \pi \varepsilon_o$.

Electric force and electric field are in the same direction for a positive charge and in opposite directions for a negative charge.

EXERCISE **11.3**

1. The electrical force between two charged objects is 0.02 N. If each object is given twice the original charge but the objects are located at the original distance from each other, what is the new force?

   A. 0.01 N
   B. 0.02 N
   C. 0.04 N
   D. 0.08 N

2. The electrical force between two charged objects is 0.020 N. If the objects are now moved four times as far apart, what is the new force?

   A. 0.0012 N
   B. 0.0025 N
   C. 0.0050 N
   D. 0.010 N

3. Four equally charged positive particles are held in position in a square arrangement, so that the sides of the square have a length R. What is the net force on an electron placed in the center of the square?

   A. $\mathbf{F} = \dfrac{3kq^2}{R^2}$

   B. $\mathbf{F} = \dfrac{4kq^2}{R^2}$

   C. $\mathbf{F} = \dfrac{\sqrt{2}kq^2}{R^2}$

   D. Zero

4. A stream of electrons is "shot" from a cathode-ray tube so that the electrons enter a uniform electric field between two horizontal charged parallel plates. Assuming the electrons don't strike either plate before exiting, describe the path the electrons take while through the field and after exiting the field. (Assume the gravitational force is negligible compared to electric forces.)

   A. linear and horizontal through the plates and linear and horizontal after leaving the plates
   B. linear and horizontal through the plates and parabolic after leaving the plates
   C. parabolic through the plates and parabolic after leaving the plates
   D. parabolic through the plates and linear after leaving the plates

Questions 5 and 6 use this diagram.

Two charged particles are held in position as shown, with a +3 μC charge at $x = 1$ m and a −4 μC charge at $x = 6$ m.

5. Determine the force exerted on the −4 μC charged object by the +3 μC charged object.

    A. 0.0043 N to the right
    B. 0.022 N to the right
    C. 0.0043 N to the left
    D. 0.022 N to the left

6. Determine the direction of the net force exerted by the two charged particles on a proton placed at $x = 3$m.

    A. to the right
    B. to the left
    C. no net force

7. An isolated point charge, Q, exerts a force F on a second charge, q, when they are a distance d apart. If Q and d are both doubled, the magnitude of force of F on q:

    A. is half is much
    B. remains the same
    C. is twice as much
    D. is four times as much

Electric forces exerted by charged objects on each other are equal in magnitude and opposite in direction (Newton's Third Law of Motion).

---

8. Consider the two positively charged objects positioned below, with one charge at $x = 2$ m, and a second charge at $x = 6$ m. Each object's charge is $+5 \times 10^{-6}$ coulomb. Calculate the electric force they exert on each other.

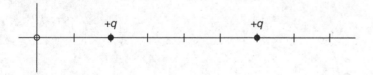

    A. 14.1 N
    B. 0.0028 N
    C. 0.056 N
    D. 0.014 N

9. What is the electric force exerted on an electron by another electron when they are at a distance of 1.0 nm apart?

    A. 0.23 MN
    B. 0.23 kN
    C. 0.23 μN
    D. 0.23 nN

Electric Potential and Potential Energy

Electric Potential, *V*: Electric potential energy per unit charge, with energy measured in joules, charge measured in coulombs, and electric potential measured in volts. Thus, 1 volt = 1 joule/coulomb.

$$V = \frac{1}{4\pi\varepsilon_o}\frac{q}{R} = \frac{kq}{R}$$

$$V = \frac{PE}{q}$$

Electric Potential Energy, *PE* or *U*: Potential energy of a charge with respect to one or more other charges. The total potential energy of a system of charges is the sum of the potential energy of each charge with respect to each other charge, taken two at a time.

$$\text{Electric potential energy} = PE_E = U_E = qV = \frac{kq_1q_2}{R}$$

EXAMPLE—Electric Equipotential Lines

In the situation shown below, an object charged with positive 2.0 μC is held in position. The dashed circles around the charge represent **equipotentials,** which are three dimensional points in space that are all the same distance from the charge and all have the same potential. Point **B** lies on an equipotential that is 2.0×10^{-4} m from the charge, and point **A** is on an equipotential is 3.0×10^{-4} m from the charge.

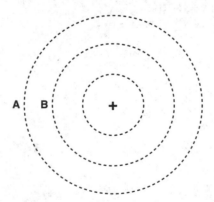

(a) Calculate the absolute potential at point **B** due to the positive charge. (b) Calculate the potential difference from point **B** to point **A**. (c) Determine the work done by the electric field in moving a proton from **B** to **A**.

Solution:

(a) $\quad V_B = kq/r = \dfrac{(9\times10^9\,\text{N}\times\text{m}^2/\text{C}^2)(2\times10^{-6}\,\text{C})}{2\times10^{-4}\,\text{m}} = 9\times10^7\,\text{v}$

(b) $V_A = kq/r = \dfrac{(9 \times 10^9 \text{N} \times \text{m}^2/\text{C}^2)(2 \times 10^{-6} \text{C})}{3 \times 10^{-4} \text{m}} = 6 \times 10^7 \text{ v}$

$\Delta V = V_f - V_o = V_A - V_B = 6 \times 10^7 - 9 \times 10^7 = -3 \times 10^7 \text{ V}$

This answer makes sense. Moving from **B** to **A**, the absolute potential decreases, so the change in absolute electric potential is negative.

(c) $W = q\Delta V = (1.6 \times 10^{-19} \text{C})(-3 \times 10^7) = -4.8 \times 10^{-12} \text{ J}$

This work is negative, since the field is causing a positive charge to move to a lower potential at **A**, decreasing its potential energy. It is much like lowering a heavy bowling ball to a lower potential energy—it is negative work.

EXAMPLE—Motion of a Charge Between Charged Plates

A proton is positioned between the charged metal plates below, near the positive plate, where it has maximum potential energy. The potential difference between the plates is 100 volts, and the distance between the plates is 1 cm. Calculate: (a) the magnitude of the uniform electric field between the plates; and (b) the velocity of the proton as it reaches the negative plate, if it is released from rest at the positive plate.

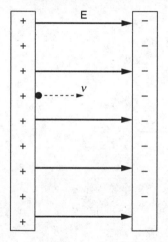

Solution:

(a) $|E| = \dfrac{V}{d} = \dfrac{100 \text{ v}}{0.01 \text{ m}} = 10{,}000 \text{ V/m}$

Notice that the positively charged proton has a force on it in the same direction as the electric field, so it "falls" from the positive plate to the negative plate.

(b)   Use conservation of energy. The electrical potential energy of the proton when it is near the positive plate becomes kinetic energy as the proton moves toward the negative plate.

$\Delta U_E = \Delta K$

$q\Delta V = \frac{1}{2}m(\Delta v)^2$

$(1.6 \times 10^{-19} \text{C})(100 \text{ v}) = \frac{1}{2}(1.67 \times 10^{-27} \text{ kg})(v^2)$

$v = 1.38 \times 10^5 \text{ m/s}$

EXAMPLE—Electric Potential Energy

Three objects, all with an equal charge of +10 μC, are placed on the corners of an equilateral triangle. The length of the each side of the triangle is 1 m. Calculate the electric potential energy of the system of three charged objects.

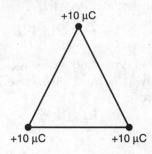

+10 μC

+10 μC   +10 μC

Solution:

The electric potential energy of a system is the sum of the potential energies between each of the charges, taken two at a time.

$$U_{total} = \frac{kq_1q_2}{R_{12}} + \frac{kq_1q_3}{R_{13}} + \frac{kq_2q_3}{R_{23}} = \frac{(9\times10^9 \, \frac{N\text{-}m^2}{C^2})(10\times10^{-6}\,C)(10\times10^{-6}\,C)}{1\,m} \times 3$$

$$U_{total} = 0.9 \, J$$

EXERCISE

11.4

Questions 1–3 use this diagram.

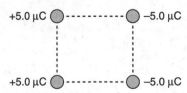

+5.0 μC ⚪----------⚪ −5.0 μC

+5.0 μC ⚪----------⚪ −5.0 μC

1. The four charges shown above are held in position on the four corners of a square that has a length of 50 cm on each side. What is the absolute potential in the center of the square?

   A. 10 V
   B. 20 V
   C. −10 V
   D. Zero

2. In the diagram above of the four charges on the corners of a square, suppose that all four charges are the same magnitude and all four charges are positive. The absolute potential at the center of the square will be:

   A. positive
   B. negative
   C. zero

3. In the diagram above for the four charges on the corners of a square, suppose all four charges are of the same magnitude but all four charges are negative. The absolute potential at the center of the square will be:

A. positive
B. negative
C. zero

Questions 4 and 5 use the following diagram.

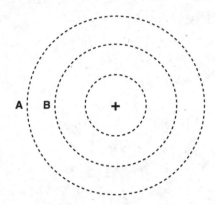

4. In the situation shown below, an object charged with positive 2.0 μC is held in position. Point **B** is 0.004 m from the charge, and point **A** is 0.008 m from the charge. Compare the absolute potential at **B** to the absolute potential at **A**.

A. Twice as much
B. Half as much
C. The same
D. Four times as much

5. An object with a charge of +5 μC is placed at point **B** in the diagram above. What is the electric potential energy of the system of two charges?

A. 90 J
B. 0.00025 J
C. 22.5 J
D. 45 J

6. A hollow metal sphere has a total charge of 10 μC on its surface. The sphere has a diameter of 2.0 m and is held in position with its center at $x = 2$ m. Determine the potential energy of the system if a proton is placed at $x = 8$ m.

A. 4.0×10^{-16} J
B. 2.4×10^{-15} J
C. 4.8×10^{-16} J
D. Zero

DC Circuits

- Electric Potential Difference and Current
- Resistance and Resistivity
- Capacitance
- Ohm's Law
- Conservation of Charge and Kirchhoff's Junction Rule
- Conservation of Energy and Kirchhoff's Loop Rule
- Power

In this chapter, you will review the basic of simple direct-current (DC) circuits, including what causes current to flow in circuits and how to analyze circuits of different configurations. The study of alternating-current (AC) circuits, which are the type used in household circuitry, is beyond an introductory course and not included in this review. Here are some symbols that will be used in diagrams of circuits—called **schematic diagrams**.

—⋀⋀⋀—	Resistor	—(A)—	Ammeter
—⊣⎮⎮⊢—	Battery	—(V)—	Voltmeter
—⊣⊢—	Capacitor	—╱—	Switch

Electric Potential Difference and Current

The previous chapter reviewed the concept of **electric potential,** which is a scalar quantity measured in volts that is positive around a positive charge and negative around a negative charge:

$$V = \frac{U_E}{q}$$

$$V = \frac{1}{4\pi\varepsilon_o}\frac{Q}{r} = \frac{kQ}{r}$$

Electric Potential Difference, ΔV: The difference, measured in volts, of electric potential between two different points. Positively charged objects move from higher electric potential to lower electric potential, and rate of flow is related to size of electric potential difference.

NGSS HS-PS2-6 and HS-PS-PS3-1 and HS-PS4-5

Ohm's Law describes a relationship between electric potential difference, current, and resistance.

$$\Delta V = IR$$

Potential difference, or **voltage**, is measured with a voltmeter, which should be connected in parallel with the circuit component or region over which potential difference is being measured. The schematic diagram below shows how to connect a voltmeter to measure potential difference across a battery.

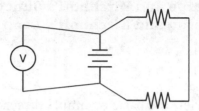

Electric Current, *I*: Rate of flow of electric charge in a circuit, measured in amperes (amps, A) or coulombs per second. The direction of **conventional current** flow in the external circuit is from higher potential (longer bar on the battery in the diagrams) to lower potential (shorter bar on the battery). Sometimes battery diagrams are labeled with + for higher potential and − for lower potential.

$$I = \frac{\Delta q}{\Delta t}$$

Electric current is measured in a circuit with an **ammeter,** which should be connected in series with the electric component through which current is being measured. The diagram below shows the correct connection of an ammeter in a series circuit and the direction of conventional current flow from the battery (high-potential side is the longer bar). Since the current is the same everywhere in the circuit, the ammeter could be connected anywhere in series between any two of the components and have the same reading.

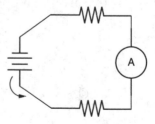

Positive charges move from higher potential to lower potential, and negative charges move from lower potential to higher potential.

1. When a net charge of 3.6 μC moves past a given point in 10 ms, what is the electric current?

   A. 0.36 mA
   B. 0.36 A
   C. 3.6 A
   D. 36 A

2. Which of the following quantities is equivalent to one ampere (A)?

   A. 1 J/s
   B. 1 J·s/C
   C. 1 C/s
   D. 1 V/s

Questions 3–4 use the following schematic diagram.

3. In the circuit diagram below, the resistors are connected to a voltage of 20 V. Assuming negligible resistance in the wires and in the battery, determine the direction of the current in the 3-ohm resistor when the switch is closed and whether the current in the 3-ohm resistor will be greater than or less than the current in the 6-ohm resistor.

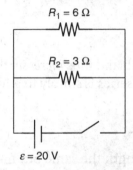

   A. Current is to the right and less than in the 6-ohm resistor.
   B. Current is to the left and less than in the 6-ohm resistor.
   C. Current is to the right and greater than in the 6-ohm resistor.
   D. Current is to the left and greater than in the 6-ohm resistor.

4. If the current from the battery is 10 A, is the current in the 3-ohm resistor equal to, less than, or equal to, less than, or greater than 10 A?

   A. The current is equal to 10 A.
   B. The current is greater than 10 A.
   C. The current is less than 10 A.

5. What is the difference between conventional current flow and electron flow?

   A. Conventional current flows from low potential to high potential.
   B. Electrons flow from low potential to high potential.
   C. Conventional current always flows clockwise around the circuits.
   D. Electrons do not actually move in the circuit.

Resistance and Resistivity

Electrical resistance is the opposition to the flow of electric current in a conductor. Resistance is measured in ohms (Ω). For an ohmic resistor, current is directly proportional to potential difference (ΔV) and inversely proportional to resistance (R), according to Ohm's law:

$$I = \frac{\Delta V}{R}$$

Electrical Conductor: A material that conducts electricity. Metals tend to be good conductors.

Electrical Insulator: A material such as glass or wood that does not conduct electricity.

Resistivity, ρ: A property of materials that defines whether or not that material is a good conductor of electricity (low value of resistivity) or a poor conductor of electricity (high value of resistivity). Resistivity increases with temperature in most materials. For example, silver is an excellent conductor of electricity and has a resistivity of 1.59×10^{-8} Ω-m. Nichrome wire is a poor conductor of electricity, with a resistivity of 100×10^{-8} Ω-m. (As a matter of fact, nichrome has such a high resistivity and resistance that it is used in toasters—where it gets hot enough to toast!)

Resistance increases with resistivity, length, and temperature of a resistor and decreases with cross-sectional area:

$$R = \frac{\rho L}{A}$$

Connecting two identical resistors in **series** effectively increases the length and thus *increases* the total resistance. The rule for connecting two or more resistors of any size in series is simply to add them:

$$R_{\text{total}} = \sum R_i$$

Connecting two identical resistors in **parallel** effectively keeps the length the same but increases the cross-sectional area for current—which *decreases* total resistance. The rule for connecting two or more resistors of any size in parallel is:

$$\frac{1}{R_{\text{total}}} = \sum \frac{1}{R_i}$$

The two resistors below are connected in parallel. The total resistance of this combination is 2 ohms.

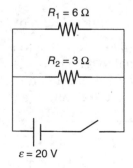

> When adding *two* resistors in parallel, a simpler method is to do "product over sum." For example, if a 3-ohm and 6-ohm resistor are in parallel, the equivalent resistance is 18/9, or 2 ohms. The interesting thing here is that adding resistors in parallel *decreases* the total resistance.

Electromotive force (EMF), ε: Sometimes used for potential difference, but means potential difference of a battery without any internal resistance. Voltage and potential difference and EMF will be used interchangeably in this text, since we will be dealing with **ideal batteries** that don't have internal resistance.

EXERCISE 12.2

Questions 1–4 use this diagram.

The circuit below consists of a 6 V battery, a switch, and three resistors, connected as shown, with the 100 Ω and the 200 Ω in series with each other and the 300 Ω in parallel with the other two.

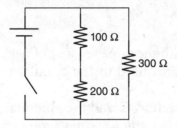

1. Determine the total resistance of the three resistors connected as shown.

   A. 600 Ω
   B. 520 Ω
   C. 150 Ω
   D. 55 Ω

2. If the 300-ohm resistor is disconnected, what is the total resistance?

   A. 100 Ω
   B. 300 Ω
   C. 67 Ω
   D. 150 Ω

3. If the 100-ohm resistor is taken out of the original circuit and the wires are reconnected, leaving the other two as they are shown, what will be the total resistance?

   A. 500 Ω
   B. 250 Ω
   C. 120 Ω
   D. 100 Ω

4. If the 200-ohm resistor is replaced by a 400-ohm resistor in the original circuit:

   A. The total resistance of the circuit is smaller.
   B. The current through the 100 Ω is the same as before.
   C. The potential difference across the 300-ohm resistor is less than before.
   D. The current in the circuit from the battery is less.

5. A 0.5-mm-diameter graphite pencil "lead" is used as a resistor. It has resistance R. What would be the predicted resistance of a second piece of "lead" that has diameter 1.0 mm and is half as long as the first piece?

   A. $R/4$
   B. $R/2$
   C. $R/8$
   D. $2R$

Capacitance

Capacitor: A device in an electrical circuit that stores energy as electrical charge. Capacitance is directly proportional to the area of the plates and inversely proportional to distance between the plates. If the capacitor contains a dielectric, the value of κ (the dielectric constant) for the dielectric is inserted.

$$C = \frac{\kappa \varepsilon_o A}{d}$$

For parallel plate capacitors, the charge is stored on metal plates, with the same amount of negative charge on one plate as positive charge on the other plate. The area used in the calculation of capacitance is the area of overlap of the two plates.

    Charge stored in a capacitor depends on the size of the capacitor, C, and the potential difference to which the capacitor is charged, V. *An uncharged capacitor still has capacitance, but it does not hold charge or energy until it is charged to a potential difference, V.*

$$Q = CV$$

Energy: The energy is stored in the electric field between the plates of a capacitor. The amount of energy stored is proportional to the size of the capacitor and the charging electric potential difference:

$$PE = \tfrac{1}{2}CV^2$$

Combining two or more capacitors:

For two or more capacitors connected in series: $\dfrac{1}{C_{\text{total}}} = \sum \dfrac{1}{C_i}$

For two or more capacitors connected in parallel: $C_{\text{total}} = \sum C_i$

> Notice that the rules for combining resistors and combining capacitors in series and in parallel are the *opposite* of each other. The reason comes from their structures: resistance increases with length and decreases with area, and capacitance increases with area and decreases with distance between plates.

Dielectrics: Insulating materials placed between the plates of capacitors to (a) prevent charge from moving directly between the plates, and (b) physically hold the capacitor plates apart so that an electric field can develop between the plates.

Dielectric constant, κ: A value unique to different materials that describes the ability of the material to serve as a dielectric; it is the ratio of the capacitance with that material as a dielectric to the same capacitor with vacuum between the plates. For example, the dielectric constant for vacuum is 1, and the value for air is close to 1. A higher dielectric constant increases the capacitance of a capacitor and allows it to store more energy. The **permittivity, ε_o,** is the value used when calculating capacitance for a any capacitor:

$$C = \frac{\kappa \varepsilon_o A}{d} \qquad \text{where } \varepsilon_o = 8.85 \times 10^{-12}\,\text{F/m}$$

Capacitance is measured in farads, F. The diagram below shows the direction of current flow as a capacitor is charged by a battery. The resistor in the circuit slows the rate of charging without affecting the amount of charge or potential difference on the fully charged capacitor. When the capacitor is fully charged, there is no current flowing in the circuit, so there is no voltage drop across the resistor. Applying Kirchhoff's loop rule (see Chapter 12), the voltage on the fully charged capacitor is the same as the voltage of the battery.

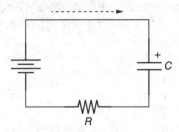

> Though capacitors and resistors in circuits add to each other in opposite ways, the conservation laws apply the same to both. The current is the same for two resistors in series in the same way that the charge is the same on two capacitors in series. And the potential difference on two equal resistors in parallel is the same in the same way that the potential difference on two capacitors in parallel is the same. In both cases, Kirchhoff's rules apply.

EXAMPLE (Advanced)—Capacitors in Parallel

In the circuit below, two identical 400-µF capacitors are connected in parallel to a 10 V power source. After the switch has been closed for a long time and the capacitors are fully charged, determine: (a) the potential difference in each capacitor; (b) the charge on each capacitor; (c) the electrical energy stored in each capacitor.

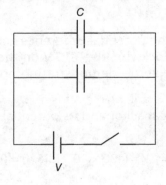

Solution:

(a) Apply Kirchhoff's loop rule to this circuit. For the small loop and for the larger loop, the potential difference on the capacitor has to the equal to the potential difference of the power supply. Therefore, the potential difference on each capacitor is 10 V.

(b) Now that we know the capacitance of each capacitor and the potential difference, apply the equation to find the charge on each capacitor:

$$Q = CV$$

$$Q = (400 \times 10^{-6} \text{ F})(10 \text{ V}) = 0.004 \text{ C}$$

(c) On each capacitor:

$$U_E = \tfrac{1}{2}CV^2 = \tfrac{1}{2}(400 \times 10^{-6} \text{ F})(10 \text{ V})^2 = 0.02 \text{ J}$$

We can check these results by adding the two capacitors (since they are in parallel) to get a total of 800 μF and then use the equation to find the total charge on both capacitors and the total energy stored on both capacitors:

$$Q_{total} = CV = (800 \times 10^{-6} \text{ F})(10 \text{ V}) = 0.008 \text{ C}$$
$$U_{total} = \tfrac{1}{2}CV^2 = \tfrac{1}{2}(800 \times 10^{-6} \text{ F})(10 \text{ V})^2 = 0.04 \text{ J}$$

Both of these values check correctly with the total calculated above for both resistors. The total charge is the sum of each, since the charge splits in the two branches, and the total energy is the sum of both.

EXAMPLE (Advanced)—Capacitors in Series

In the circuit below, two identical 400-μF capacitors are connected in series to a 10 V power source. After the switch has been closed for a long time and the capacitors are fully charged, determine: (a) the potential difference in each capacitor; (b) the charge on each capacitor; (c) the electrical energy stored in each capacitor.

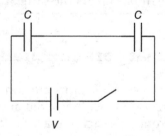

Solution:

(a) Using Kirchhoff's loop rule, the potential difference across each capacitor must be 5 V. As we go around the loop clockwise, the +10 V on the power supply adds to the two −5 V on the capacitors to be zero—demonstrating no net energy change in the loop and conserving energy.

(b) Use the capacitance of each capacitor and its potential difference to find the charge on each capacitor.

$$Q = CV = (400 \times 10^{-6} \text{ F})(5 \text{ V}) = 0.002 \text{ C}$$

(c) $U_E = \frac{1}{2}CV^2 = \frac{1}{2}(400 \times 10^{-6})(5\text{ V})^2 = 0.005\text{ J}$

We can check these results by adding the two capacitors using "product over sum" (since they are in series and add in the same way two resistors in parallel would add) to get a total of 200 μF and then use the equations to find the total charge on both capacitors and the total energy stored on both capacitors:

$$Q_{total} = CV = (200 \times 10^{-6}\text{ F})(10\text{ V}) = 0.002\text{ C}$$
$$U_{total} = \frac{1}{2}CV^2 = \frac{1}{2}(200 \times 10^{-6}\text{ F})(10\text{ V})^2 = 0.01\text{ J}$$

Both of these values check correctly with the total calculated above for both resistors. Charge is conserved; since the capacitors are in series, the charge deposited on the positive plate of one causes an equal amount of charge to be removed from the negative plate and then to be deposited on the positive plate of the next capacitor plate. The total energy is equal to the sum of the energies on the two capacitors.

EXAMPLE (Advanced)—Charging a Capacitor

In the circuit below, a 100-mF capacitor is connected in series with a 1,000-ohm resistor and a 10 V power supply. The capacitor is uncharged prior to closing the switch. (a) What is the current through the resistor immediately after the switch is closed? (b) What is the current through the resistor after the switch has been closed for a very long time? (c) What are the potential difference and charge on the capacitor after the capacitor is fully charged?

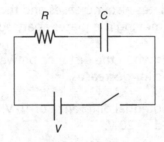

Solution:

(a) The capacitor allows no current to flow through, and initially the capacitor is uncharged, so all the potential difference in the external circuit is through the resistor. Thus by Kirchhoff's loop rule, the potential difference through the resistor equals the potential difference supplied, or 10 V.

(b) After a very long time, the capacitor is fully charged to 10 V. Throughout the charging process, as the potential difference between the positive plate of the battery and the positive plate of the capacitor become closer in value, the current decreases. So there is no current through the resistor after a long time.

(c) The capacitor is fully charged at 10 V, so the energy in the capacitor is:

$$U_E = \frac{1}{2}CV^2 = \frac{1}{2}(1,000 \times 10^{-6}\text{ F})(10\text{ V})^2 = 0.05\text{ J}$$

1. A parallel plate capacitor stores 100 μJ of energy when charged to an electric potential difference of 20 V. If the capacitor is charged again to a potential difference of only 10 V, what will be the energy stored in the same capacitor?

   A. 200 μJ
   B. 100 μJ
   C. 50 μJ
   D. 25 μJ

2. A 2,000-μF capacitor is charged to 10 V. What is the energy stored in the capacitor?

   A. 0.2 J
   B. 0.1 J
   C. 200,000 J
   D. 100,000 J

3. A 2,000-μF capacitor is charged to 10 V. Without removing any of the charges or changing the plate area or dielectric, the plates are moved apart so that there is twice the distance between them as before. What happens to the energy stored in the capacitor?

   A. Positive work is done in moving the plates apart, so the energy stored in the capacitor increases.
   B. The capacitance increases when the distance between plates increases, so the potential difference between the plates also increases and the energy in the capacitor increases.
   C. Negative work is done in moving the plates apart, so the energy stored in the capacitor decreases.
   D. The capacitance decreases when the distance between the plates increases, so the energy stored in the capacitor decreases.

4. A capacitor is charged to a potential difference of 10 V. Compare the charge stored in the same capacitor if it is charged to 20 V.

   A. The charge is the same in both cases.
   B. There is one half as much charge at 20 V.
   C. There is twice as much charge at 20 V.
   D. There is four times as much charge at 20 V.

5. A 2,000-μF capacitor and a 4,000-μF capacitor are connected in series in a circuit with a 100 V battery. What is the equivalent capacitance of the two capacitors?

   A. 1,300 μF
   B. 2,000 μF
   C. 3,000 μF
   D. 6,000 μF

6. A 2,000-μF capacitor and a 4,000-μF capacitor are connected in parallel in a circuit with a 100 V battery. What is the equivalent capacitance of the two capacitors?

   A. 6,000 μF
   B. 3,000 μF
   C. 2,000 μF
   D. 1,300 μF

Ohm's Law

Ohm's Law describes the relationship between the **potential difference, ΔV**, the **current, I**, and **resistance, R** in a circuit or in a circuit element or any part of a circuit. It is written here in the form where we can see that the amount of current is directly proportional to potential difference and inversely proportional to resistance.

$$I = \frac{\Delta V}{R}$$

Electric Current: Rate of flow of charge in a **conductor**, measured in amperes or amps (A). The direction of **conventional current** flow in physics is defined as the direction of positive charge flow, which is from higher potential to lower potential.

Potential difference: Difference in electric potential (measured in volts, V) between one point and another. A potential difference, ΔV, is necessary for charge to flow.

Resistance: R, measured in ohms (Ω) reduces the rate of flow of charge (or decreases current) in a circuit.

EXERCISE
12.4

Questions 1–3 refer to this diagram.

The circuit below consists of a 6 V battery, a switch, and three resistors, connected as shown.

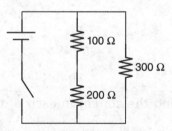

1. Determine the current from the battery.

   A. 0.01 A
   B. 0.02 A
   C. 0.03 A
   D. 0.04 A

2. Rank the resistors in order of the amount of current flowing through each.

   A. 100 Ω > 200 Ω > 300 Ω
   B. 300 Ω > 200 Ω > 100 Ω
   C. 100 Ω = 300 Ω > 200 Ω
   D. 100 Ω = 300 Ω = 200 Ω

3. In the above circuit, determine the current through the 300 Ω resistor.

   A. 33 mA upward on the page
   B. 16 mA downward on the page
   C. 20 mA downward on the page
   D. 40 mA downward on the page

4. In a circuit, a 20-ohm resistor and a 30-ohm resistor are connected in series to a 10 V battery. After the switch is closed, what is the current in the 20-ohm resistor?

    A. 5 A
    B. 0.4 A
    C. 0.1 A
    D. 0.2 A

5. Students in the laboratory find data for the current through a resistor as they change the potential difference. They plot potential difference versus current. The slope of their graph line is:

    A. power consumed by the resistor
    B. energy used by the resistor
    C. resistance of the resistor
    D. resistivity of the resistor

Conservation of Charge and Kirchhoff's Junction Rule

Kirchhoff's Junction Rule: A statement of conservation of charge. At any point in a circuit, the total current flowing into that point is equal to the total current flowing out. As charge flows through a circuit element, the energy of the charge may be converted to other forms such as heat or light, but the charge itself is not created or destroyed:

$$\Sigma I_{In} = \Sigma I_{Out}$$

EXAMPLE—Combination Series/Parallel Circuit

In the circuit below, the battery potential difference is 12 V. Find the current in each of the resistors.

Solution:

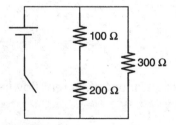

Find the total resistance of the three resistors first. Add the two resistors in series in the middle branch to get 300 ohms. Then use "product over sum" with the 300 ohms in each of the two parallel branches: $R = (300)(300)/600$, which is 90,000/600 or 150 Ω.
    Use Ohm's law with $\Delta V = 12$ V and $R = 150$ Ω. $I = 12/150 = 0.08$ A.
    Kirchhoff's Loop Rule says that the current leaving the battery splits when it reaches the junction at the parallel branches. In this case, the resistance of each branch is the same, so the current splits equally. The 300-ohm resistor gets 0.04 A and the other branch gets 0.04 A.

The current must be the same through both resistors in that branch, so the 100-ohm resistor gets 0.04 A and the 200-ohm resistor gets 0.04.

The currents from the branches recombine at the second junction so the amount of current returning to the battery is the same as the current that left the battery.

---

1. This is a simple series circuit with two resistors and a battery and a switch, with the switch closed. An ammeter is correctly connected in series to measure current. If the ammeter is moved to a position between the high potential side of the battery and the resistor next to it, how will the ammeter reading change?

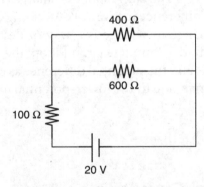

   A. The ammeter reading will be less.
   B. The ammeter reading will be greater.
   C. The ammeter reading will be the same.

Questions 2–5 use this diagram.

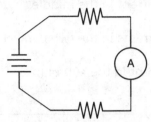

2. In the circuit above, what are the current leaving the battery and the current in the 100-ohm resistor?

   A. 0.04 A in both
   B. 0.06 A in both
   C. 0.06 A leaving the battery and 0.03 A in the 100-ohm resistor
   D. 0.04 A leaving the battery and 0.02 A in the 100-ohm resistor

3. What happens to the current when it reaches the junction to the parallel branches?

   A. It splits so half goes through each branch.
   B. The current is the same in each branch as it was through the 100-ohm resistor.
   C. 2/5 of the current goes through the top branch and 3/5 goes through the middle branch.
   D. 3/5 of the current goes through the top branch and 2/5 goes through the middle branch.

4. Compare the current from the high-potential side of the battery and going through the 100-ohm resistor to the current returning to the low potential side of the battery.

   A. They are the same.
   B. The current leaving the battery is more than the current returning to the battery.
   C. The current leaving the battery is less than the current returning to the battery.

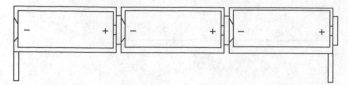

5. Which statement supports Kirchhoff's junction rule?

   A. The current leaving the battery is the same as the current flowing in the top branch.
   B. The current in the top parallel branch must be equal to the current in the middle branch.
   C. The current leaving the battery must equal the sum of the currents in the two parallel branches.
   D. The current in the 100-ohm resistor must be equal to the current in the 400-ohm resistor.

Conservation of Energy and Kirchhoff's Loop Rule

Kirchhoff's Loop Rule: A statement of conservation of energy: The sum of all the changes in voltage around a complete circuit loop is zero. Electric potential difference is a way of describing change in energy per charge. Therefore, potential difference or change in voltage is a way of describing change in energy. Following conventional current flow around a complete circuit loop, the potential difference as current moves through a resistor is negative, and the potential difference as current moves through the battery (from the high-potential or positive side to the lower-potential or negative side) is positive:

$$\sum \Delta V_{\text{Loop}} = 0$$

EXAMPLE—Combination Series and Parallel Circuit

For the circuit below, determine: (a) the total resistance; (b) current through the 100-ohm resistor; (c) current through the 400-ohm resistor; (d) potential difference across the 600-ohm resistor; and (e) total power consumption by the circuit.

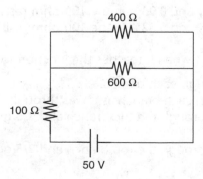

Solution:

Step 1. To find total resistance, start with the two resistors in parallel, by doing "product over sum." Then add the resistor that is in series.

$$R_p = \frac{(400)(600)}{(400+600)}\Omega = 240\ \Omega$$
$$R_T = 100 + 240 = 340\ \Omega$$

Step 2. All current must flow through the 100-ohm resistor, so calculate the total current from the battery, using Ohm's Law and the battery voltage and the total resistance:

$$V = IR$$
$$50\ V = (I)(340\ \Omega)$$
$$I = 0.15\ A$$

This is the current from the battery and also through the 100-ohm resistor.

Step 3. We'll approach this using the loop rule. First, calculate the potential difference across the 100-ohm resistor: $V = IR = (0.15\ A)(100\ \Omega) = 15\ V$
Now apply the loop rule to the large loop containing the battery and the 100-ohm resistor and the 400-ohm resistor. Add the potential difference for each element in the loop. Start by going through the battery (+50 W), then through the 100-ohm resistor (−15 V), leaving −35 V for the 400-ohm resistor so that the total is zero. If $V = 35\ V$ and $R = 400\ \Omega$, then $I = V/R = 0.088\ A$.

Step 4. If we use the loop rule again, except going through the smaller loop with the 600-ohm resistor, the potential difference has to be the same for that part of the loop as it was for the 400-ohm part of the larger loop. Therefore, the potential difference across the 600-ohm resistor is 35 V.

Step 5. Use the battery output power: $P = VI = (50\ V)(0.15\ A) = 7.5\ W$
If you calculate the power consumption of each resistor separately, you find that the total of the three will also be 7.5 W.

EXERCISE 12.6

1. This is a simple series circuit with two identical resistors and a battery and a switch, with the switch closed. A voltmeter is correctly connected in parallel with the battery to measure the potential difference (voltage) across the battery. How does the potential difference across the battery compare to the potential difference across one of the resistors?

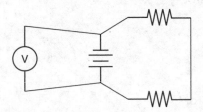

   A. It is half as much.
   B. It is twice as much.
   C. It is the same.

2. Below is a simple series circuit with two identical resistors and a battery and a switch, with the switch closed. A voltmeter is correctly connected in parallel with one of the resistors to measure the potential difference (voltage) across that resistor. The voltmeter reading is 4 V. What would be the voltmeter reading if the voltmeter is connected in parallel with the battery instead?

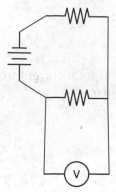

A. 4 V

B. 8 V

C. 12 V

D. Zero. The voltmeter should not be connected in parallel.

Questions 3–5 use the circuit below.

The total resistance of all three resistors is 340 Ω and the current from the battery is 0.3 A.

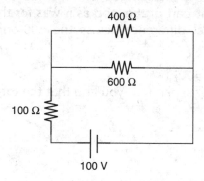

3. What is the potential difference across the 100 ohm resistor?

A. 100 V

B. 30 V

C. 70 V

D. Zero

4. What is the potential difference across the 400-ohm resistor?

A. 100 V minus the ΔV across the 100-ohm resistor

B. 100 V

C. 50 V

D. Half the ΔV across the 100-ohm resistor

5. What is the potential difference across the 600 ohm resistor?

A. the total of the voltage across the 100-ohm and the 400-ohm resistors

B. the same as the battery voltage

C. the same as across the 100-ohm resistor

D. the same as across the 400-ohm resistor

Power

Electrical power is the rate at which energy is used or dissipated, measured in joules per second, or **watts**. The total power used by the components of a circuit is equal to the power output by the battery.

$$P = \frac{\Delta E}{\Delta t}$$

Electrical power in a circuit (in watts) is calculated with the following formula:

$$P = VI \qquad \text{or} \qquad P = I^2R \qquad \text{or} \qquad P = \frac{V^2}{R}$$

Electrical Energy is equal to power times time. The energy is in joules if power is in watts and time is in seconds. However, for household purposes, the power is often measured in kilowatts and the time in hours, so the energy usage is measured in kilowatt-hours (kWh).

> Energy can be measured in *joules* (J) or *kilowatt-hours* (kWh) or in *electron volts* (eV).

EXAMPLE—Current and Power

Three identical resistors, each 1 MΩ, are connected as shown below to a potential difference of 10 V. Determine the current in the resistor in the top branch and the power consumed by that resistor.

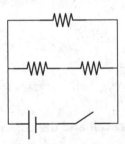

Solution:

The resistor on the top of the diagram is wired in parallel with the other two resistors, which are wired in series with each other.

Step 1. Find the total resistance in the circuit. Add the two resistance values in the middle branch, since they are in series, to find the 2-MΩ resistance of the two in series. Then combine that 2-MΩ resistance, using the parallel rule, with the 1-MΩ resistance in the top branch.

$$\frac{1}{R} = \frac{1}{R_1} + \frac{1}{R_2} = \frac{1}{1\,\text{M}\Omega} + \frac{1}{2\,\text{M}\Omega}$$
$$R = 0.67\,\text{M}\Omega$$

or

$$R = \frac{\text{product}}{\text{sum}} = \frac{(1 \times 10^6\,\Omega)(2 \times 10^6\,\Omega)}{3 \times 10^6\,\Omega} = \frac{2 \times 10^{12}\,\Omega}{3 \times 10^6\,\Omega} = 0.67\,\text{M}\Omega$$

Step 2. Find the current coming from the battery.

$$I = \frac{\Delta V}{R} = \frac{10\ V}{0.67\ M\Omega} = 1.5 \times 10^{-5}\ A$$

Step 3. Find the current in that top resistor. The current will split in the two branches, with 2/3 of the current going into the top branch and 1/3 of the resistance in the branch with two resistors.

The current in that top branch is $(2/3)(1.5 \times 10^{-5}\ A)$, which is $1 \times 10^{-5}\ A$.

Another way to do this is to use Kirchhoff's loop rule. Make a loop with the battery and the top resistor, so the voltage goes up by 10 V in the battery and down by 10 V in the resistor. We now know that resistor has $V = 10\ V$ and $R = 1\ M\Omega$, so $I = V/R = 1 \times 10^{-5}\ A$.

Step 4. Calculate the power in that resistor: $P = VI = (10\ V)\ (1 \times 10^{-5}\ A) = 1 \times 10^{-4}\ W$

EXERCISE
12.7

Questions 1–3 use the following circuit.

The circuit has three identical 100-ohm resistors connected to a 10 V battery

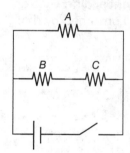

1. What is the total resistance in the circuit and the current leaving the battery?

   A. 0.033 A
   B. 0.15 A
   C. 0.30 A
   D. 0.60 A

2. What is the power produced by the battery?

   A. 0.15 W
   B. 0.30 W
   C. 1.5 W
   D. 3.0 W

3. Rank the energy used by each of the resistors in the circuit, from greatest to least:

   A. A > B > C
   B. A > B < C
   C. A > B = C
   D. A = B = C

4. The electrical output to a laptop computer is labeled 19.5 V and 3.34 A. Determine the energy used by the computer during 30 minutes of recharging.

   A. 175 J
   B. 1,950 J
   C. 1.2×10^5 J
   D. 0.036 J

5. A 40 W light bulb is burned for about 60 hours per month. How much energy does it use in a month?

   A. 2.4 kWh
   B. 8,640 kWh
   C. 2,400 kWh
   D. 0.024 kWh

Magnetism

- **Magnetic Fields**
- **Magnetic Forces**
- **Electromagnetism**

The link between moving charges and changing magnetic fields is the focus of this chapter: charged particles moving in magnetic fields experience magnetic forces; current-carrying wires are surrounded by magnetic fields; changing magnetic fields induce currents in wires. Electromagnetic induction can be a difficult topic, so alternate solutions to these problems will, hopefully, provide you with at least one "sure" method of solution.

Magnetic Fields

Magnetic Field, *B*: The region in space surrounding a magnetic field that exerts a magnetic force on another magnetic field or on a moving object with electric charge; measured in teslas (T). Magnetic field has a direction from what we call "north" to what we call "south." Magnetic field lines are continuous. A current-carrying wire has a magnetic field in the region around the wire that is in a circular pattern defined by using a right-hand rule: Point the thumb in the direction of the current (or direction of movement of positively charged particles) and curl the finger around the wire in the direction of the magnetic field. Magnetic field strength increases with the electrical current and decreases with distance from the wire.

$$B = \frac{\mu_0 I}{2\pi r}$$

EXERCISE

13.1

1. A wire carries a current to the right on the page. Where could the magnetic field around the wire be directed out of the page?

   A. above the wire
   B. below the wire
   C. along the wire, in the same direction the current is flowing
   D. along the wire, in the opposite direction the current is flowing

NGSS HS-PS2-1

2. Calculate the strength of the magnetic field at a distance of 0.2 m from a wire carrying a current of 3 A.

   A. 3×10^{-7}
   B. 3×10^{-6}
   C. 6×10^{-7}
   D. $3\pi \times 10^{-7}$

3. The magnetic field is measured at point P at a distance r from a current carrying wire. If the current in the wire is doubled, the magnetic field strength at point P is:

   A. Half as much
   B. One-quarter as much
   C. twice as much
   D. four times as much

4. The magnetic field is measured at point P at a distance r from a current carrying wire. If the distance from the wire is doubled, the magnetic field strength at point P is:

   A. Half as much
   B. One-quarter as much
   C. twice as much
   D. four times as much

5. The magnetic field is measured at point P at a distance r from a current carrying wire. If the length of the wire is doubled, the magnetic field strength at point P is:

   A. Half as much
   B. One-quarter as much
   C. twice as much
   D. the same

Magnetic Forces

Magnetic Force: Measured in newtons, exerted between two magnets or exerted by a magnetic field on an object with charge in motion within the magnetic field or on a current-carrying wire in a magnetic field.

Force on a charge moving in a magnetic field:

$$\overline{F}_B = q\overline{v} \times \overline{B} = q\overline{v}\overline{B}\sin\theta$$

Force on a current-carrying wire in a magnetic field:

$$\overline{F}_B = IL_\perp B$$

Right Hand Rule: A convention used to determine the direction of a vector cross product. In the context of magnetism in this chapter, it's used to determine the direction of the magnetic force $\overline{F}_B$ exerted on an object with electric charge q moving with velocity $\overline{v}$ through a magnetic field $\overline{B}$. The equation is: $\overline{F}_B = q\overline{v} \times \overline{B}$. The order in which the multiplication takes place is essential, in the same way that the direction of the x-axis crossed with y-axis produces the z-axis: $\overline{x} \times \overline{y} = \overline{z}$.

To use the right-hand rule, use the index finger as the first vector to be multiplied. The other fingers are the second vector and must be directed perpendicular to the index finger. The product, which is also a vector, is in the direction of the thumb—and must be directed perpendicular to the other two directions. It is important to do the operation in this order. You can test this with 3-d axes: x crossed with y is equal to z.

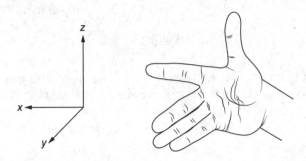

Students often forget to switch the direction of the force when using the right-hand rule for negatively charged particles moving in a magnetic field. So here's a somewhat controversial suggestion: Use your left hand when using the "right-hand rule" for negative particles. It works! Just be consistent about it.

EXAMPLE—Charge Moving Between Charged Plates

An electric field of strength 800 N/C is set up between two charged parallel plates as shown. An electron enters the field from the left at a speed of 6,000 m/s. (a) What direction is the electric field between the plates? (b) What is the electric force on the electron as it enters the field? (c) Determine the magnitude and size of magnetic field that could be set up between the plates so that the electron moves between the plates with no net force.

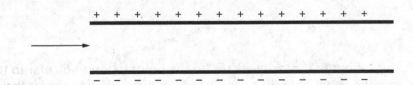

Solution:

(a) The electric field is directed from the positive plate to the negative plate, or downward on the page.

(b) First, determine the magnitude of the electric force:

$$|\mathbf{F}| = qE = (1.6 \times 10^{-19} \text{ C})(800 \text{ N/C}) = 1.28 \times 10^{-16} \text{ N}$$

The force on a negative charge is in the opposite direction of the field, so a force upward on the page will be exerted on the electron by the electric field.

(c) For there to be no net force on the electron, the magnetic field introduced between the plates must exert a magnetic force downward on the page. The magnitude of the magnetic force must be equal to the magnitude of the electric force, so $\mathbf{F}_B = 1.28 \times 10^{-16}$ N. Now find the strength of the magnetic field:

$$\left|\mathbf{F}_B\right| = q\mathbf{v}\mathbf{B}$$

$$1.28 \times 10^{-16}\ \text{N} = (1.6 \times 10^{-19}\,\text{C})(6{,}000\ \text{m/s})(\mathbf{B})$$

$$\left|\mathbf{B}\right| = 0.13\ \text{T}$$

Using the right-hand rule, with velocity to the right and force downward on the page, (remembering to use the left hand since it's a negative particle), we see that the magnetic field must be directed into the page.

EXAMPLE—Charge Moving in a Magnetic Field

An electron is moving at a velocity of 6.0×10^5 m/s to the right on the page as it enters a uniform magnetic field of strength 100 T directed into the page, as shown below. (a) Determine the magnitude and direction of the force exerted on the electron as it initially enters the field, and discuss the path the electron will take if it remains in the field for a long time. (b) Determine the force and direction on a proton entering the field at the same speed in the same direction.

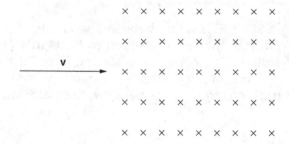

Solution:

(a) First, we recognize that the velocity vector (to the right) is perpendicular to the magnetic field vector (into the page). Using the right-hand rule, with index finger for the velocity vector and other fingers for the magnetic field (which is into the page), the thumb indicates direction of the magnetic force, which is toward the top of the page. *However*, since the electron is a negative charge, we have to reverse that direction, so our answer is: *toward the bottom of the page.*

We don't have vector components to be concerned about here, since all the vectors are perpendicular, so the magnitude of the force is determined by multiplication.

$$\left|\mathbf{F}\right| = \left|q\mathbf{v}\mathbf{B}\right| = (1.6 \times 10^{-19}\ \text{C})(6 \times 10^6\ \text{m/s})(100\ \text{T}) = 9.6 \times 10^{-11}\ \text{N}$$

If the field is expansive enough that the electron stays in the field, the direction of the velocity changes as the force is exerted on the electron, so the electron moves in a circle, with the magnetic force exerting the centripetal force.

(b) If the moving object is a proton instead, the only part of the answer that changes is the direction. Since mass does not play a role in the calculation, the larger mass of the proton does not affect the size of the force. The proton has exactly the same charge as the electron, so the magnitude of the force is 9.6×10^{-11} N. The difference is in the direction of the force. Using the right-hand rule, we see that the force on the proton as it enters the field is toward the top of the page.

EXAMPLE—Magnetic Force as the Centripetal Force

A proton enters a magnetic field with a strength of 100 T at speed $v = 1 \times 10^6$ m/s. The field is large enough in area that the proton becomes confined to the field. (a) Describe the motion of the proton in the field. (b) Calculate the radius of the proton's motion in the field.

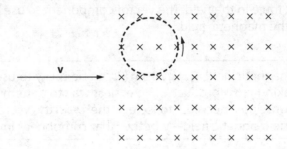

Solution:

(a) When the proton initially enters the field, the magnetic force is upward on the page. However, this changes the proton's path, causing it to continually change direction toward the center of a counterclockwise circle, with the magnetic force providing the centripetal force for the proton. The magnetic force has a constant value, so the centripetal acceleration can be calculated. The radius of the subsequent circular path, r, is constant and is called the **cyclotron radius.**

(b) Set the magnetic force on the proton as the centripetal force and solve for r:

$$q\mathbf{v}\mathbf{B} = \frac{mv^2}{r}$$

$$r = \frac{m\mathbf{v}}{q\mathbf{B}} = \frac{(1.67 \times 10^{-27} \text{ kg})(1 \times 10^6 \text{m/s})}{(1.6 \times 10^{-19} \text{ C})(100 \text{ T})} = 1.04 \times 10^{-4} \text{m}$$

EXAMPLE (Advanced)—Charge Moving at an Angle into a Magnetic Field

A proton is moving at a velocity of 6.0×10^5 m/s in the direction shown below as it enters a uniform magnetic field of strength 100 T directed toward the bottom of the page. (a) Determine the magnitude and direction of the force exerted on the proton as it initially enters the field, and (b) discuss the path the proton will take if it remains in the field for a long time.

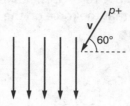

Solution:

(a) Since the velocity vector is not perpendicular to the magnetic field vectors, we must recognize that only the component of the velocity that is perpendicular to the field will produce a magnetic force in the field. The vector component to use here is <u>v cos 60°</u>, since it is perpendicular to the magnetic field.

> The equation for magnetic field, $\mathbf{F} = q\mathbf{v}B \sin \theta$, can be tricky to use, since most students are not adept at taking cross products. The easiest way to determine the correct answer is to make the conscious decision to select the velocity vector component that is perpendicular to the magnetic field—whether that component involves sine or cosine.

Now use the equation to calculate the magnitude of the magnetic force:

$$|\mathbf{F}| = |q\mathbf{v}B| = (1.6 \times 10^{-19} \text{ C})(6 \times 10^6 \text{ m/s cos } 60°)(100 \text{ T}) = 4.8 \times 10^{-11} \text{ N}$$

Use the right-hand rule to determine the direction of the magnetic force, with the correct component of velocity (to the left) as the index finger, the magnetic field (toward the bottom of the page) as the other fingers, and the force *out of the page*. This is a positive particle, so the direction is correct as determined by the right hand.

(b) The perpendicular component of the velocity (v cos 60°) produces a force in the magnetic field so that the proton moves in a circle in and out of the page while it's in the field. The magnetic force provides the centripetal force, so the radius of the circle could be calculated by setting the magnetic force equal to the expression for centripetal force.

$$F_{centripetal} = F_{magnetic}$$
$$\frac{m\mathbf{v}^2}{r} = q(\mathbf{v} \cos 60°)\mathbf{B}$$

The other component of the velocity (v sin 60°) is the inertial path along the field lines for the proton. This component produces no force but keeps the proton moving downward on the page. As a result, the proton circles in and out of the page as it moves downward on the page, producing a spiral path.

Questions 1 and 2 refer to this diagram.

1. An electron with velocity $v = 2.0 \times 10^5$ m/s moves into a uniform magnetic field, $B = 200$ T. Determine the magnitude and direction of the magnetic force on the particle when it first enters the field.

   A. 400 N out of the page
   B. 6×10^{-12} N out of the page
   C. 400 N into the page
   D. 6×10^{-12} N into the page

2. Suppose the electron in the diagram above approaches the field instead at an angle (toward the upper right). Which is true, compared to the above situation?

   A. The magnetic force on the electron will have the same magnitude as before.
   B. The magnetic force will be in the same direction as before.
   C. The magnetic force will be downward on the page.
   D. There will be no magnetic force on the electron.

3. A proton moving in the same direction as magnetic field lines:

   A. experiences a magnetic force in the opposite of that experienced by an electron
   B. experiences a magnetic force perpendicular to its direction of motion
   C. experiences a magnetic force in the same direction as its motion
   D. experiences no magnetic force

4. In which case will the greatest force be exerted on the charged particle?

   A. a proton moving at 60,000 m/s through an electric field of strength 400 N/m
   B. a stationary electron in an electric field of strength 800 N/m
   C. a proton moving at 60,000 m/s perpendicularly through a magnetic field of strength 400 T
   D. a stationary proton in a magnetic field of strength 800 T

   > Charged particles must be moving to experience a force in a magnetic field.

5. What is the direction of the magnetic force exerted by a magnetic field of strength 80 T on an electron placed at rest in the center of it?

   A. upward on the page
   B. downward on the page
   C. out of the page
   D. no force

6. What is the direction of the electric force exerted by an electric field of 80 N/C downward on the page on an electron placed at rest in the center of it?

   A. upward on the page
   B. downward on the page
   C. out of the page
   D. no force

7. A proton moves from the left into the uniform magnetic field directed to the right as shown below. What is the direction of the magnetic force on the proton?

   A. upward on the page
   B. downward on the page
   C. out of the page
   D. no force

Questions 8–10 use the diagram below.

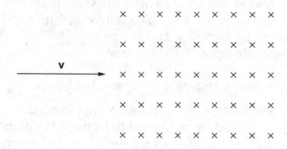

8. A proton is moving to the right on the page as it enters a uniform magnetic field directed into the page, as shown above. Determine the direction of the force exerted on the electron as it initially enters the field.

   A. upward on the page
   B. downward on the page
   C. to the right
   D. into the page

9. If the magnetic field is very large in area and a proton enters the field perpendicular to the magnetic field lines, what type of path will the proton take after it has been in the field a long time?

   A. parabolic
   B. circular
   C. linear
   D. helical

10. An electric field could also be applied to the situation shown in Question 8 so that the proton goes straight to the right without changing its path due to the magnetic field. What direction should the electric field be directed?

   A. upward on the page
   B. downward on the page
   C. out of the page
   D. down into the page

Electromagnetism

Electromagnetism: Describes the interactions of electricity and magnetism. In this unit, we will review how magnetic fields can induce currents in wires moving through the fields, how current carrying wires can create magnetic fields, and how current-carrying wires can exert forces on each other.

EXAMPLE—Forces Due to Current-Carrying Wires

Two current-carrying wires are held a distance 0.2 m apart, as shown, with point P halfway between them. The current in wire 1 is 1.0 A in the positive x-direction and the current in wire 2 is 1.0 A, also in the positive x-direction.

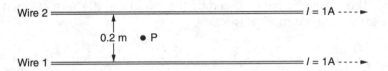

(a) Determine the magnitude and direction of the magnetic field created at point P by wire 2.

(b) Determine the magnitude and direction of the net magnetic field created at point P by both wires.

(c) Assuming the lengths of wire examined here are each 1.5 m in length, determine the magnitude and direction of the force on wire 2 created by wire 1.

(d) Determine the magnitude and direction of the force on wire 1 created by wire 2.

Solution:

(a) Using the right-hand rule with thumb in direction of current and fingers curled around the wire, the **B** field due to wire 2 at point P is into the page. Use the equation to calculate magnitude:

$$\mathbf{B} = \frac{\mu_o I}{2\pi R} = \frac{(4\pi \times 10^{-7})(1\,\text{A})}{(2\pi)(0.1\,\text{m})} = 2 \times 10^{-6}\,\text{T}$$

(b) Wire 1 is creating a magnetic field at point P that has the same magnitude (since all the numbers are the same in the equation), except wire 1 creates a field out of the page. Therefore, the net magnetic field due to both wires is zero.

(c) To calculate the magnetic field that wire 1 creates at wire 2 due to the 0.2 m distance between them, calculate the same as in (b) but double the distance from 0.1 m to 0.2 m, so **B** = 1 × 10⁻⁶ T. Use the equation **F** = *ILB* and then use the right-hand rule with index finger along the direction of current, other fingers along the magnetic field direction, and thumb toward force. The force is downward on the page.

$$\mathbf{F} = IL_{\perp}\,\mathbf{B} = (1\ \text{A})(1.5\ \text{m})(1 \times 10^{-6}\ \text{T}) = 1.5 \times 10^{-6}\ \text{N}$$

(d) The magnetic field is the same and length of wire is the same and current is the same, so **F** is the same. The field created by wire 2 is the same direction at wire 1 as it is at point P—into the page.

Using the right-hand rule again, this time we see that the force is upward on the page.

EXAMPLE—Magnetic Field Around a Current-Carrying Wire

A straight wire is carrying a current of 2.0 A in the positive *x*-direction. Determine the magnitude and direction of the magnetic field created by the current in the wire at point P, which is located 0.1 m in the positive *y*-direction from the wire.

Solution:

The magnitude of the magnetic field at point P is calculated from the equation:

$$B = \frac{\mu_0 I}{2\pi r} = \frac{(4\pi \times 10^{-7}\ \text{T·m/A})(2\ \text{A})}{2\pi(0.1\ \text{m})}$$

$$B = 4 \times 10^{-6}\ \text{T}$$

To find the direction of the magnetic field at point P, use a variation of the right-hand rule. Curl the fingers of your right hand and stick out your thumb. Point your thumb in the direction of the current in the wire. Your fingers curl in the pattern of the magnetic field, which circles the wire. At P, the magnetic field will be directed out of the page.

EXAMPLE—Forces Between Current-Carrying Wires

Two current-carrying wires are held a distance 0.2 m apart, as shown, with point P halfway between them. The current in wire 1 is 2.0 A in the positive *x*-direction and the current in wire 2 is 3.0 A in the negative *x*-direction.

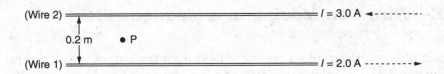

(a) Determine the magnitude and direction of the magnetic field created at point P by wire 2.

(b) Determine the magnitude and direction of the <u>net magnetic field</u> created at point P by both wires.

(c) Assuming the lengths of wire examined here are each 1.5 m in length, determine the magnitude and direction of the force on wire 1 created by wire 2.

(d) Determine the magnitude and direction of the force on wire 1 created by wire 2.

Solution:

(a) Use the same equation and methods as described in SP3, using the current and distance given.

$$\mathbf{B} = \frac{\mu_0 I}{2\pi r} = \frac{(4\pi \times 10^{-7}\, \text{T·m/A})(3\,\text{A})}{2\pi(0.1\,\text{m})}$$

$$\mathbf{B} = 6 \times 10^{-6}\,\text{T}$$

Using the right-hand rule as described in SP3, the magnetic field created at point P by the current in wire 2 is out of the page.

(b) First determine the magnetic field created by wire 1 at point P.

$$\mathbf{B} = \frac{\mu_0 I}{2\pi r} = \frac{(4\pi \times 10^{-7}\, \text{T·m/A})(2\,\text{A})}{2\pi(0.1\,\text{m})}$$

$$\mathbf{B} = 4 \times 10^{-6}\,\text{T}$$

Using the right-hand rule, the direction of this field at point P is also out of the page.

Since the fields from both wires are out of the page, the two vectors add to produce a net magnetic field of 1×10^{-5} T out of the page.

(c) The magnitude of the magnetic force on wire 1 created by the magnetic field of wire 2 depends on the size of the field from wire 2 at wire 1 (a distance of 0.2 m), which must be calculated first. The force then depends on that field from wire 2 and the current in wire.

$$|\mathbf{F}| = I_2 L B_1 = I_2 L \left(\frac{\mu_0 I_1}{2\pi r} \right)$$

$$|\mathbf{F}| = (3\,\text{A})(1.5\,\text{m}) \left[\frac{(4\pi \times 10^{-7}\, \text{T} \times \text{m/A})(2\text{A})}{2\pi(0.2\,\text{m})} \right] = 9 10^{-6}\,\text{N}$$

(d) The forces are equal in magnitude and opposite in direction.

EXAMPLE—Wire Coil Moving in a Magnetic Field

A 1-m² loop of wire moves into a magnetic field from the left, as shown below, at a speed of 30 m/s. The wire has a resistance of 1 MΩ, and the magnetic field has a strength of 100 T.

(a) What is the current induced in the wire when the loop is halfway into the field?

(b) What is the current induced in the wire when the loop is still moving but is completely inside the field?

(c) What is the current induced in the wire when the loop has moved halfway out of the field?

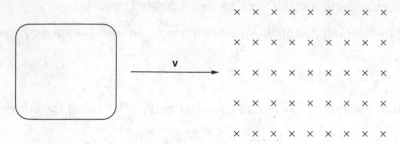

Solution:

(a) As the loop enters the field, the only net force is exerted by the field in the front edge of the wire.

There are a couple of ways to determine the direction of current for a wire loop moving in a magnetic field:

(1) Consider a single, positively charged particle in the leading edge of the wire and use the right-hand rule, with that particle moving to the right in a field into the page. The force is upward on the page, so the current flows counterclockwise in the wire.

(2) Using Lenz's Law, the magnetic flux increases into the page through the loop as the loop enters the field. Thus, a current will be induced in the loop that creates an increasing magnetic field out of the page to counteract that change. Using the right hand, curly our fingers counterclockwise around the loop so your thumb points outward from the page to show the magnetic field outward created by that current.

To find the size of the current, first calculate the magnitude of EMF induced:

$$\varepsilon = \frac{\Delta f}{\Delta t} = \frac{\Delta(\mathbf{B}A)}{\Delta t} = \frac{\mathbf{B}\Delta(lw)}{\Delta t} = \mathbf{B}lv, \text{ where } l \text{ is the length of the end piece of wire}$$

$$\varepsilon = (100 \text{ T})(1 \text{ m})(30 \text{ m/s}) = 3,000 \text{ V}$$

Now use Ohm's Law to determine the current. $I = \frac{\varepsilon}{R} = \frac{3,000 \text{ V}}{10^6 \text{ } \Omega} = 0.003 \text{ A}$

(b) When the loop is completely inside the field, there is no change in magnetic flux through the loop, so there is no current induced.

(c) As the loop leaves the field at the same speed, the EMF and current induced are the same size but opposite in direction. Check this with the right-hand rule again, using one positively charged object in the wire remaining in the field. The velocity is to the right and field is into the page, so the force is upward on the page, causing current to flow clockwise in the loop.

1. A current-carrying loop of wire is placed in a magnetic field. In which case will the field exert the largest torque on the wire?

   A. The plane of the loop is parallel to the field lines.
   B. The plane of the loop is perpendicular to the field lines.
   C. The plane of the loop is at a 45° angle to the field lines.
   D. The plane of the loop is at a 60° angle to the field lines.

Questions 2–5 use this diagram.

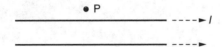

Two long, parallel wires each carry a current of 0.5 A to the right. Point P, above, is located 0.2 cm above the top wire and 0.4 cm above the bottom wire.

2. The direction of the magnetic field at point P due to the top wire is:

   A. into the page
   B. out of the page
   C. to the right
   D. zero

3. The force the top wire exerts on each meter of the bottom wire is:

   A. 1.2×10^{-5} N upward on the page
   B. 1.2×10^{-5} N downward on the page
   C. 2.5×10^{-5} N upward on the page
   D. 2.5×10^{-5} N downward on the page

4. Determine the direction of the net magnetic field at point P due to the two wires.

   A. to the right
   B. to the left
   C. into the page
   D. out of the page

5. If the current in the top wire is reversed, what is the direction of the net magnetic field at point P due to the two wires?

   A. to the right
   B. to the left
   C. into the page
   D. out of the page

Physical and Geometric Optics

> - **Electromagnetic Waves**
> - **Reflection and Mirrors**
> - **Refraction and Lenses**
> - **Interference and Diffraction**

Light, to a physicist, is more than the small range of visible frequencies we as humans are able to see. The study of light includes all the properties and behaviors of a wide range of electromagnetic frequencies and wavelengths—from gamma radiation at the high-energy/high-frequency end of the spectrum to radio waves at the low-energy/low-frequency end.

This chapter includes some simple rules and guidelines for construction of ray diagrams in geometric optics and detailed sample problems demonstrating how to solve problems with diffraction and thin-film interference, though these last topics are often not included in introductory courses.

Electromagnetic Waves

Electromagnetic radiation (often referred by physicists generally as *light*): a form of energy that is transmitted through space. In a vacuum, light travels at a constant speed, 3×10^8 m/s.

Electromagnetic Spectrum: The range of frequencies and wavelengths for electromagnetic radiation.

Type of Radiation	Frequency Range (Hz)	Wavelength Range (m)
Radio/TV	~10^8	1–100
Microwave	~10^{10}	0.001–1
Infrared	~10^{12}–10^{14}	10^{-6}–10^{-4}
Visible	~10^{14}	(400–700 nm)
Ultraviolet	~10^{15}–10^{17}	10^{-7}–10^{-8}
X-rays	~10^{17}–10^{20}	10^{-9}–10^{-12}
Gamma rays	~10^{20}–10^{22}	$<10^{-12}$

Frequency, f: The number of oscillations per second, measured in hertz (Hz). Frequency defines the color of visible light and is used to assign names to ranges of other forms of electromagnetic radiation.

Wavelength, λ: The distance from a point on one wave to a comparable point on the next wave (e.g., crest to crest), measured in meters.

> Light is visible to humans in the wavelength range of 400 to 700 nanometers (nm), with 400 nm at the red end of the visible spectrum and 700 nm at the violet end of the spectrum.

Speed of light, c: Speed of electromagnetic wave propagation in a vacuum, which is 3.0×10^8 m/s. In a vacuum, $c = f\lambda$.

Wave speed, v: Speed of propagation of an electromagnetic wave in a medium, $v = f\lambda$. The speed of light in a vacuum is c, and the speed of light in another medium with **index of refraction, n,** is equal to **c/n.**

Monochromatic Light: Light that consists of waves with a single frequency and therefore a single color. Laser light is monochromatic and coherent.

Intensity: Power per unit of area (watts per square meter) of energy given off from a source, which decreases as the inverse square of the distance from the source; often referred to as "brightness."

Polarization: A wave property of light that occurs when the electric field is restricted to one plane. As shown here, two crossed polarizer sheets can block light completely. When light passes through a polarizer, the intensity of the polarized light is less than the intensity of the incident light.

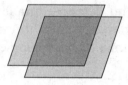

Rays: Vectors that indicate the direction in which the light is propagated. These rays are drawn perpendicular to the wave fronts. (Read more about Huygens to learn more about wave fronts.) When light rays enter a new medium, the **law of refraction** (or Snell's Law) describes the direction of those rays:

$$n_1 \sin\theta_1 = n_2 \sin\theta_2$$

Dispersion: Each color component of white light refracts at a different angle, separating white light into colors. As the white light strikes this prism face at an angle to the normal, dispersion occurs. The violet end of the spectrum refracts at the largest angle, and red refracts least. The other colors appear in order in between.

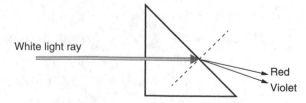

Normal: A line constructed perpendicular to the surface. In the diagram above, the white light enters along the normal, so it does not change direction as it passes into the prism. No dispersion occurs here. However, when light strikes a normal at an angle, it will reflect and/or refract at an angle to the normal.

Color Mixing: When all the frequencies of visible light are mixed in equal proportions, the light appears white. Black is the absence of light (except for a few reflected rays to make a surface visible. When red and blue light mix, humans see magenta. When red and green light mix, humans see yellow. And when green and blue light mix, we see cyan.

Color filters: Filters reflect or absorb some frequencies and allow others to pass through. A magenta surface, for example, will reflect red and blue light. A red shirt reflects red light and absorbs all other colors. A red filter allows red light to pass through, so if a red filter is used to view a red shirt, the shirt will appear red. However, if a blue filter is used to view a red shirt, the shirt will appear to be black.

EXERCISE 14.1

1. The type or "color" of electromagnetic radiation we detect depends upon its:

   A. wavelength
   B. speed
   C. frequency
   D. polarization

2. An object that looks blue when viewed under a white light source will appear to be what color when illuminated by white light and viewed through a yellow filter?

   A. orange
   B. black
   C. red
   D. green

3. Which answer choice correctly ranks these radiations from lowest frequency to highest?

   A. red, green, infrared, gamma
   B. infrared, blue, ultraviolet, x-ray
   C. yellow, red, infrared, radio
   D. ultraviolet, green, red, infrared

   > Note: The Doppler shift for light is not calculated using the same formula as the Doppler shift for sound. However, the "red shift" of light does indicate that the emitter of the light is traveling away and frequency is lowered.

4. Wavelengths for visible light ranges from about 400 to 700 nm. What is the visible frequency range?

   A. 4.3×10^5 Hz to 7.5×10^5 Hz
   B. 2.4×10^5 Hz to 5.3×10^5 Hz
   C. 4.3×10^{14} Hz to 7.5×10^{14} Hz
   D. 4.3×10^{15} Hz to 7.5×10^{15} Hz

5. Your television remote control likely operates by emitting a wavelength of about 1,000 nm. What type of light does the control emit?

   A. blue light
   B. infrared
   C. red light
   D. ultraviolet

6. Determine the wavelength of light in a bubble film, if the wavelength of that light is 624 nm in air. The index of refraction of the film is 1.3.

   A. 480 nm
   B. 811 nm
   C. 370 nm
   D. The same as in air

7. A ray of light goes from air with $n = 1$ into glass with $n = 1.5$. Which statement is true about what happens to the light when it enters the glass?

   A. Its wavelength and frequency both decrease.
   B. Its wavelength and frequency both increase.
   C. Its wavelength decreases and frequency stays the same.
   D. Its wavelength stays the same and frequency decreases.

8. A white shirt is illuminated by full spectrum white light. If the shirt is viewed through overlapping magenta and yellow filters, the shirt will appear to be:

   A. black
   B. yellow
   C. red
   D. magenta

9. What is the frequency of a green laser light with a wavelength of 532 nm?

   A. 7.82×10^9 Hz
   B. 5.66×10^{14} Hz
   C. 6.02×10^{23} Hz
   D. 4.14×10^{14} Hz

Reflection and Mirrors

Convex Mirror: The middle of the mirrored side curves outward; does not form real images, so all images are upright and virtual.

Concave Mirror: The middle of the mirrored side "caves" inward; forms an upright, enlarged, virtual image when the object is inside the focal point; forms an inverted, real image when the object is outside the focal point.

Plane Mirror: Flat mirror that produces an upright (virtual) image that appears the same distance behind the mirror as the object is positioned in front of the mirror. This image is reversed left to right.

Law of Reflection: The incident angle and reflected angles are equal when measured to the normal.

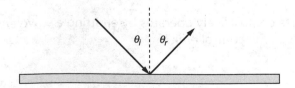

Real Image: Can be projected on a screen and is inverted. Produced by thin lenses and mirrors when: (1) the object is placed outside the focal length of a convex lens, or (2) the object is placed outside the focal length of a concave mirror.

Virtual Image: Cannot be projected on a screen; appears upright.

Focal Length, *f*: A distance that is one-half the **radius of curvature, *C*,** for a circular lens or mirror; also the distance from a convex lens where transmitted parallel light rays converge or the distance from a concave mirror where reflected parallel light rays converge.

Principal Axis: A line drawn perpendicular to the center of a curved lens or mirror, so it also lies along the radius curvature. The focal point is on the principal axis. For purposes of ray drawing, the principal axis defines major rays for construction, and the object is drawn with its base on the principal axis.

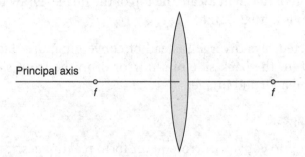

Thin lens or mirror equation: Relates focal length of a thin lens or mirror to the distance of the object from the lens or mirror (d_o) and distance of the image from the lens or mirror (d_i).

$$\frac{1}{f} = \frac{1}{d_o} + \frac{1}{d_i}$$

When using the formula, convex lenses and concave mirrors are assigned positive focal lengths, and convex mirrors and concave lenses are assigned negative focal lengths. All measurements can be made in either meters or centimeters—as long as units are consistent.

Magnification: The ratio of image height to object height or ratio of image distance to object distance:

$$M = \frac{d_i}{d_o} = \frac{h_i}{h_o}$$

Guidelines for Construction of Ray Diagrams for Geometric Optics:

1. Draw a principal axis through the center of the lens or mirror.

2. Mark focal lengths, if given.

3. Set the object on the principal axis at the specified distance (to scale), with the *bottom* of the object on the axis. Now we know that the bottom of the image will be on the axis also. The task is just to locate the top image to construct the image.

4. Select enough rays to construct so that the intersection of rays can locate the top of the image.

   - A ray constructed through the top of the object and parallel to the principal axis will be directed toward the focal point on the other side of the lens (or reflected back through the focal point on the same side of a mirror).

   - A ray constructed through the focal point and toward the top of the object as it is directed toward the lens or mirror will go exit a lens parallel to the principal axis (or be reflected parallel to the principal axis for a mirror).

   - A ray that goes through the top of the object toward the center of the lens will go straight through the center to the other side (or be reflected at the same angle back to the same side of a mirror).

5. The intersection of rays will locate the top of the image. Draw the image perpendicular to the principal axis at that point.

6. If the constructed rays diverge instead of converging, use dashed lines to project those rays backward (to the back side of a mirror or to the front side of a lens) until they do converge to find a virtual image.

General Rules for Images:

1. Concave lenses and convex mirrors never form real images.

2. Concave mirrors and convex lenses form real images only if the object is placed outside the focal length.

3. Real images form on the opposite side of a lens from the object and form on the same side of a mirror as the object. They form from light rays as they reflect and refract.

4. Virtual images are formed from the projection of light rays. They form on the "back" side of mirrors and form on the same side of a lens as the object and do not form from actual light rays.

5. Real images are inverted and virtual images are upright.

6. Plane mirrors form virtual images that appear to be the same distance behind the mirror as the distance of the object from the front of the mirror.

7. Images in plane mirrors appear left-right inverted.

EXAMPLE—Concave Mirror

An object is placed in front of a concave mirror so that it forms an image. Draw a ray diagram and describe the image.

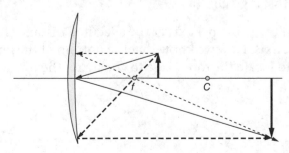

Solution:

The object is shown positioned on the principal axis and located just outside the focal point.

Ray 1 is drawn from the top of the object parallel to the principal axis, so it reflects through the focal point.

Ray 2 is drawn from the top of the object to the center of the mirror, where the principal axis meets the mirror. The reflected ray is drawn at an equal angle below the principal axis.

Ray 3 is drawn from the top of the object through the focal point. It reflects parallel to the principal axis.

All rays that go from the bottom of the object to the mirror reflect straight back along the principal axis.

The bottom of the image is located on the principal axis, and the top of the image is located where the rays reflected from the top of the object meet.

Since the image is inverted and is formed from actual reflected rays, it is a *real* image. (All real images are inverted.)

EXAMPLE—Convex Mirror

A 10-cm-tall object is placed a distance of 10 cm in front of a convex mirror with a radius of curvature of 16 cm. Find the properties of the image.

Solution:

Step 1. Ray Drawing.

(a) Set a focal point at half the radius of curvature, or at 8 cm, for purposes of construction.

(b) Draw a ray from the top of the object toward the mirror that is parallel to the principal axis, which will reflect as if it came from the direction of the focal point.

(c) Draw a ray from the top of the object parallel to the principal axis toward the focal point; it will reflect parallel.

(d) Draw a ray from the top of the object toward the center of the mirror; it will reflect at the same angle on the other side of the principal axis.

(e) Since the reflected rays do not converge, extend them backwards—behind the mirror—to locate the virtual image (upright).

Note: An image doesn't really form. A person must look into the mirror for the brain to interpret the directed rays as coming from an image behind the mirror (which, of course, is not really there).

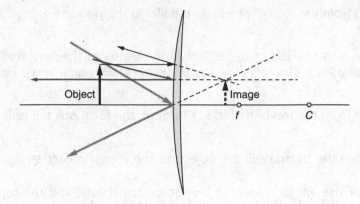

Step 2. Calculate, using the thin lens equation. (Notice that the focal length goes into the equation as a negative value, since a convex mirror does not focus light.)

$$\frac{1}{f} = \frac{1}{d_o} + \frac{1}{d_i}$$

$$\frac{1}{-8 \text{ cm}} = \frac{1}{10 \text{ cm}} + \frac{1}{d_i}$$

$$d_i = -4.4 \text{ cm}$$

This negative image distance indicates the object is behind the mirror, so it is not real. Since the image is virtual, it is upright. The ratio of image distance to object is 4.4:10, so the image is 44% as tall as the object. The image is 4.4 cm tall.

EXERCISE

14.2

1. Describe the image formed when an object is placed 5 cm in front of a convex mirror that has a radius of curvature of 20 cm.

   A. virtual, upright, larger than the object
   B. real, inverted, twice as large as the object
   C. virtual, upright, smaller than the object
   D. real, inverted, half the height of the object

Questions 2–4 use this diagram.

An object with a height of 5 cm is placed 8 cm from the center of a concave mirror that has a focal length of 6 cm. Calculate the properties of the image formed (upright or inverted, height, distance from mirror, magnification) to answer the following questions.

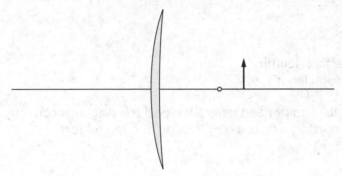

2. The image will be:

   A. upright and real
   B. upright and virtual
   C. inverted and real
   D. inverted and virtual

3. The distance of the image from the mirror is:

   A. 8 cm
   B. 6 cm
   C. 24 cm
   D. 1/24 cm

4. The magnification of the image is:

   A. ¼
   B. 3×
   C. 4×
   D. same size as object

5. When you are standing 2 m in front of a plane mirror, describe your image.

   A. same size and appears to be 2 m behind the mirror
   B. ½ size and appears 2 m in front of mirror
   C. same size and half the distance behind the mirror as you are in front of the mirror
   D. same size and at the mirror

6. The image in a plane mirror:

   A. is always real and upright
   B. is real, with a magnification of 1
   C. is virtual, with a magnification of 1
   D. is real and inverted left to right

7. Determine the type of image and magnification for an object placed 3 cm in front of a concave mirror with focal length 6 cm.

   A. virtual image, twice as large as object
   B. virtual image, half as large as object
   C. real image, twice as large as object
   D. real image, half as large as object

8. Convex mirrors form real images:

   A. never
   B. only if the object is outside the focal length
   C. only if the object is inside the focal length

9. Concave mirrors form real images:

   A. never
   B. only if the object is outside the focal length
   C. only if the object is inside the focal length

10. Light rays from the right strike a plane mirror and reflect. Which if the diagrams correctly shows the projected virtual rays a person would see by looking into the mirror?

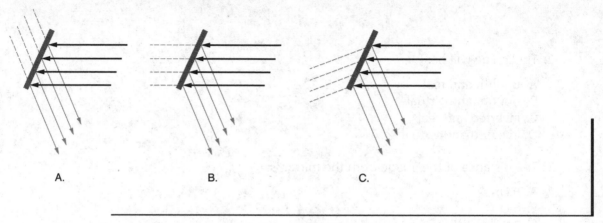

A.   B.   C.

Refraction and Lenses

Refraction: Change in speed and wavelength of light as it travels from one type of medium to another, with no change in frequency. A change in direction also results if the light does not enter the new medium along the normal. All forms of electromagnetic radiation travel at the speed of light ($c = 3 \times 10^8$ m/s) in a vacuum. When electromagnetic radiation (usually referred to as "light") enters another medium, the speed and wavelength decrease, but the frequency remains constant. Since the frequency identifies the "color," the type of radiation remains unchanged even though its speed and wavelength change. Each "color" of radiation has a different index of refraction in a given medium. Index of refraction is not strictly proportional to mass density of the medium. We often use $n = 1$ for air, since the value is closer to 1.0003 and we tend to use fewer significant digits in our calculations.

This change in speed and wavelength is called **refraction,** and the ratio of the speed in a vacuum, c, to the new speed, v, is called the **index of refraction, n.**

$$n = \frac{c}{v}$$

When light passes from a medium with lower index of refraction (such as air) into a medium with higher index of refraction (such as water or glass), its speed decreases and its wavelength decreases but its *frequency stays the same*. (Here's how to test yourself: When light shines into water, does it appear to be a different color? No. The light waves do slow down, so wavelength decreases, but frequency stays the same.)

Snell's Law describes the relationship between the angle of incidence when light strikes a boundary between two materials with different index of refraction and the angle of refraction:

$$n_i \sin\theta_i = n_R \sin\theta_R$$

For light traveling from one medium into another, this equation describes the angle of a light ray relative to the **normal** entering the medium (**incident ray**) and the angle of the ray relative to the normal as it crosses the interface (**refracted ray**) and enters the medium. Since the index of refraction of difference frequencies (colors) of light varies, each color of light will refract at a slightly different angle.

Critical Angle: Incident angle in a medium that produces an angle of refraction angle of 90°; only occurs if the incident medium has a higher index of refraction than the refractive medium. At angles larger than the critical angle in this situation, light is reflected into the medium; this is called **total internal reflection,** which is important in fiber optics.

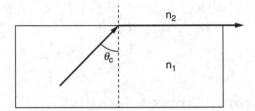

Thin Film Interference: Constructive and destructive interference of the reflected light from the top interface with the light reflecting from the bottom interface of a very thin layer of material (such as soap bubbles and thin layers of oil). This film interference can be used to block certain colors or allow certain colors, depending on the thickness of the film.

Index of Refraction, *n*: The ratio of the speed of light in a vacuum to the speed of that light in a medium. Each "color" of radiation has a different index of refraction in a given medium. Index of refraction is not strictly proportional to mass density of the medium. We often use $n = 1$ for air, since the value is closer to 1.0003 and we tend to use fewer significant digits in our calculations.

$$n = \frac{c}{v}$$

As shown above, the incident beam from the air changes direction at the interface as the light moves into the glass. Since glass has a higher index of refraction than air, the beam bends toward the normal (at a smaller angle).

Convex Lens: Curved so that it is thicker in the middle than it is on the ends; also called a converging lens, since it is able to cause parallel light rays to intersect at a focal point. As with a mirror, this focal point is one-half the radius of curvature. The magnification of a convex lens is greater when the lens has a shorter focal length—or when the lens is thicker. A convex lens will form either a real or a virtual image, depending upon where the object is placed relative to the focal length.

Concave Lens: Curved so that it is thinner in the middle; also called a diverging lens. A concave lens cannot focus light, so it does not have a real focal length and does not form a real image.

Thin-lens or mirror equation: Relates focal length of a thin lens or mirror to the distance of the object from the lens or mirror (d_o) and distance of the image from the lens or mirror (d_i):

$$\frac{1}{f} = \frac{1}{d_o} + \frac{1}{d_i}$$

Concave mirrors and convex lenses have positive focal lengths, and convex mirrors and concave lenses are assigned negative focal lengths. All measurements can be made in either meters or centimeters—as long as units are consistent.

Magnification: The ratio of image height to object height or ratio of image distance to object distance:

$$M = \frac{d_i}{d_o} = \frac{h_i}{h_o}$$

EXAMPLE—Ray Diagram, Convex Lens

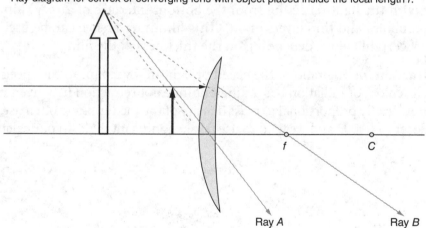

Ray diagram for convex or converging lens with object placed inside the focal length f.

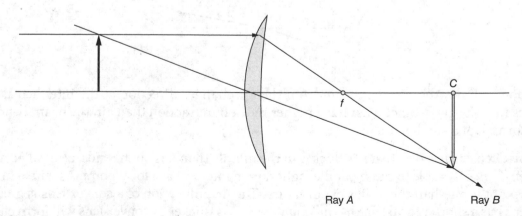

Ray diagram for convex or converging lens with object placed outside the focal length f.

Use the same rules to draw the real image, whether the object is located between the focal point and center of curvature ($C = 2f$) or outside the center of curvature. That position will only determine the size of the real image.

EXAMPLE—Convex Lens

An object is sitting on the principal axis at the center of curvature of a convex (converging) lens. Construct the ray diagram and describe the image.

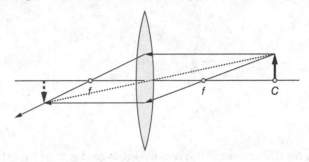

Solution:

By definition, the center of curvature is twice the focal length, so the object is at a distance of $2f$.

Ray 1 is drawn from the top of the object parallel to the principal axis so it refracts through the focal point on the far side of the lens.

Ray 2 is drawn from the top of the object through the center of the lens, so it is extended through the beyond the lens.

Ray 3 is drawn from the top of the object through the focal point on the near side of the lens, so it refracts through the lens parallel to the principal axis.

All the rays from the top of the object converge to form the top of the image, and all the rays from the bottom of the object (which is on the principal axis), extend straight through the lens to from the bottom of the image.

The image is inverted and is formed from actual rays, so the image is **real** and the same size as the object.

The size of the image can be proved using the thin lens equation:

$$\frac{1}{f} = \frac{1}{d_o} + \frac{1}{d_i}$$

$$\frac{1}{f} = \frac{1}{2f} + \frac{1}{d_i}$$

$$\frac{1}{d_i} = \frac{1}{f} - \frac{1}{2f} = \frac{1}{2f}$$

Therefore, the image distance and the object distance are the same. The magnification is equal to image distance divided by object distance, and that is 1. The image and object are the same size and are the same distance from the lens on opposite sides.

EXAMPLE—Experiment Image Formation

In a common lab experiment, a burning candle is placed on one side of a convex lens and a screen is placed on the other side of the lens so that an image forms on the screen. The candle is a distance of 10 cm from the center of a convex lens, as shown, and the screen shows a clear image of the candle when the distance from the lens to the screen is 8.0 cm. (a) Is the image on the screen real or virtual, and is the image upright or inverted? (b) What is the focal length of the lens?

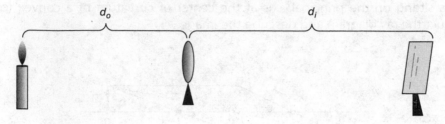

Solution:

(a) If the image shows on the screen, it is a real image—and is thus inverted with respect to the object, as are all real images.

(b) Use the thin lens equation, with object distance 10 cm and image distance 8.0 cm to solve for focal length, f:

$$\frac{1}{f} = \frac{1}{d_o} + \frac{1}{d_i}$$

$$\frac{1}{f} = \frac{1}{10} + \frac{1}{8}$$

$$\frac{1}{f} = \frac{4}{40} + \frac{5}{40} = \frac{9}{40}$$

$$f = \frac{40}{9} = 4.4 \text{ cm}$$

The focal length turns out to be positive, which is no surprise, since this is a convex or converging lens. This means the image is real, inverted, and located on the opposite side of the lens and the object.

EXAMPLE—Refraction and Snell's Law

Green light with a wavelength of 530 nm in air goes from air into a glass block at an incident angle of 40°, as shown below. Calculate: (a) the refraction angle as the light enters the block; and (b) the wavelength of the light in the block. (c) If the block is a perfectly shaped rectangular solid, calculate the angle as the light refracts back into air on the other side of the block.

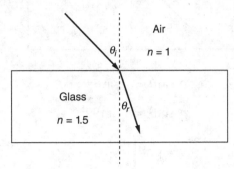

(a)   Snell's law:

$$n_i \sin \theta_i = n_R \sin \theta_R$$

$$(1.0)(\sin 40°) = (1.5)(\sin \theta_2)$$

$$\theta_R = \text{inv} \sin \left(\frac{(1.0)(\sin 40°)}{1.5} \right) = 25.4°$$

(b)

$$n = \frac{c}{v} = \frac{\lambda_{air}}{\lambda_{glass}}$$

$$1.5 = \frac{530 \, nm}{\lambda_{glass}} \qquad \lambda_{glass} = 353 \, nm$$

Notice that since $v = f\lambda$ and frequency stays constant, the speed of the light is reduced as it enters the glass with higher optical density, so the wavelength also decreases.

(c)   Since the opposite faces of the glass are parallel, the ray in the glass strikes the normal on the other side at the same angle, so we just reverse Snell's law to find the angle of the ray that exits the glass. It is 40 degrees, so rays entering and exiting opposite sides of a material where sides are parallel will be parallel to each other.

EXAMPLE—Critical Angle

Determine the critical angle for a beam of light shining from under water into air.

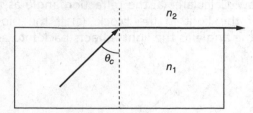

$$n_1 \sin\theta_1 = n_2 \sin\theta_2$$

$$n_1 \sin\theta_c = n_2 \sin 90°$$

$$\theta_c = \sin^{-1}\left(\frac{n_2}{n_1}\right)$$

$$\theta_c = \sin^{-1}\left(\frac{1.00}{1.33}\right) = 48.8°$$

EXAMPLE—Refraction in a Glass Block

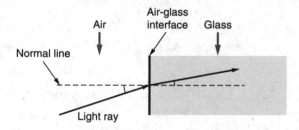

Light with a wavelength of 530 nm in air shines on a flight surface of a rectangular piece of glass. (a) If the angle of incidence is 40°, what is the angle of refraction in the glass? (b) What properties of the light change as the light enters the glass? (c) At what angle will the ray exit the glass on the other side?

Solution:

(a) Use Snell's law:

$$n_1 \sin\theta_1 = n_2 \sin\theta_2$$
$$(1.0)\sin 40° = (1.4)\sin\theta_2$$
$$\theta_2 = 27.3°$$

(b) The frequency does not change as refraction occurs. The velocity decreases as the light goes from air to glass in proportion to the wavelength decrease, since $v = f\lambda$. The index of refraction defines the velocity decrease, so it also defines the wavelength decrease.

$$n = \frac{c}{v} = \frac{\lambda_{air}}{\lambda_{glass}}$$

$$1.4 = \frac{530 \text{ nm}}{\lambda_{glass}}$$

$$\lambda_{glass} = 379 \text{ nm}$$

(c) Since the sides of the block are parallel, the Snell's law equation is just reversed at the second boundary. The exit angle will be 40°. The exit beam will be parallel to the incident beam on the first surface.

EXAMPLE—Convex Lens Magnification

An object that is 10 cm tall is placed on the principal axis a distance of 6 cm from a convex mirror. The focal length of the mirror is 4 cm. Draw a ray diagram to show where an image forms, and calculate the type, position, and magnification of the image.

Solution:

A convex lens can focus light, so f is positive. The object is outside the focal length, so we predict that a real image will form.

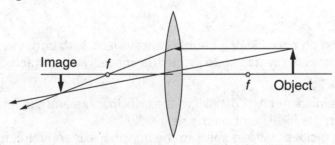

Step 1. Ray Drawing

The two simplest rays to draw are the "parallel" ray and the "center" ray. (1) A ray drawn from the top of the object parallel to the principal axis will exit the lens through the focal length on the other side. (2) A ray drawn from the top of the object through the center of the lens will pass straight through the center. The intersection of these two rays locates the top of the object—and we know the bottom of the object is on the principal axis.

We can see that the image is inverted, so it is a real image. Use calculations to confirm.

Step 2. Calculation. Using the then lens equation:

$$\frac{1}{f}=\frac{1}{d_o}+\frac{1}{d_i}$$

$$\frac{1}{4\ \text{cm}}=\frac{1}{6\ \text{cm}}+\frac{1}{d_i}$$

$$d_i=12\ \text{cm}$$

The image distance d_i is 12 cm and positive, so it is real. The ratio of image distance to object distance is 2:1, so the image is magnified 2:1 and must be 20 cm tall.

EXERCISE

14.3

Questions 1–4 refer to the diagram below.

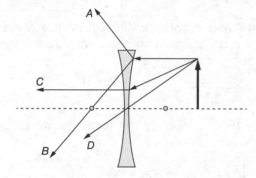

1. An object is placed on one side of a double concave lens, with both virtual focal points shown. In the process of ray tracing to locate the image, which refracted ray is <u>not</u> correct, and why?

   A. A, because the incident ray is parallel to the principal axis and the refracted ray exits as if directed from the focal point on the near side.
   B. B, because the incident ray is parallel to the principal axis and refracts toward the far focal point
   C. C, because it enters at a point above the principal axis the exits parallel to the principle axis.
   D. D, because it enters directed to the center of the mirror and exits along the same line

2. The size of the image will be:

   A. Larger than the object and located on the opposite side of the mirror from the object
   B. Smaller than the object and located on the same side of the mirror as the object
   C. Larger than the object and located on the same side of the mirror as the object
   D. Smaller than the object and located on the opposite side of the mirror from the object

3. The image formed in this case is:

   A. Real and right side up
   B. Real and inverted
   C. Virtual and right side up
   D. Virtual and inverted

4. Ray C is correct because it:

A. Hits the lens as if it came from the near-side focal point and refracts parallel to the principal axis
B. Hits the lens as if it were directed toward the far side focal point and refracts parallel to the principal axis
C. Hits the lens at the center and continues through in a straight line
D. Hits the lens parallel to the principal axis and refracts toward the far side focal point

5. If an optical medium has an index of refraction of 1.5, then we can conclude that light traveling in the medium after entering from air:

A. has two-thirds the frequency in the new medium as it had in air
B. has two-thirds the speed in the new medium as it had in air
C. must have changed its direction when entering the new medium from air
D. has 1.5 times the wavelength in the new medium as it had in air

6. When white light travels from air into a glass prism and is dispersed into colors:

A. Blue light refracts over a smaller angle than red.
B. All frequencies of light travel at the same speed.
C. Blue light refracts over a larger angle than red.

7. Determine the critical angle for plastic in air, if the plastic has an index of refraction of 1.42.

A. 36.1°
B. 44.8°
C. 86.2°
D. 52.2°

8. Convex lenses form real images:

A. never
B. only if the object is outside the focal length
C. only if the object is inside the focal length

9. Concave lenses form real images:

A. never
B. only if the object is outside the focal length
C. only if the object is inside the focal length

10. Students gather data for angle of incidence and angle of refraction for a convex lens in air. The slope of which graph would be the closest approximation of the index of refraction of the lens?

A. Sine incident angle *versus* sine refracted angle
B. Sine refracted angle *versus* sine incident angle
C. Incident angle *versus* refracted angle
D. Refracted angle *versus* incident angle

11. An upright object is placed a distance of 10 cm from a convex lens, and a real image forms a distance of 10 cm on the other side of the lens. What is the focal length of the lens?

A. 20 cm
B. 15 cm
C. 10 cm
D. 5 cm

12. Which of the ray diagrams below would be a correct ray tracing for a concave lens, with light rays entering from the right?

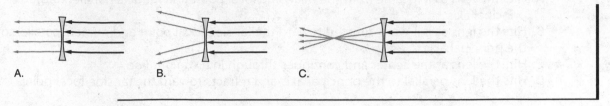

A.

B.

C.

Interference and Diffraction

Wave Interference: When waves overlap each other, moving in the same or the opposite direction, their amplitudes add to produce a new wave. If positive amplitudes or amplitudes in one direction add to negative amplitudes or amplitudes in the other direction, the result is **destructive interference.** When amplitudes add to produce larger amplitudes or reinforce the waves, it is called **constructive interference**.

Diffraction: Wave fronts passing through a small opening or around an edge may bend and interfere with each other to produce a pattern.

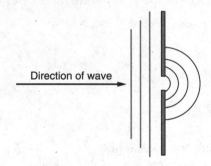

Direction of wave

EXAMPLE (Advanced)—Diffraction

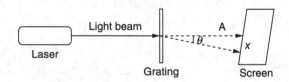

Laser

Light beam

Grating

A

θ

x

Screen

Calculate the wavelength of the laser in the following diffraction setup. The distance from the screen to the laser is 1.0 m, and the distance x from the central maximum to the first bright line is 54 cm. The diffraction grating is labeled "750 lines/mm."

m = order number (center dot is $m = 0$, and dots on either side are numbered)

λ = laser wavelength

d = slit width or slit spacing

θ = angle of beam from central maximum for the given order bright line

x = distance from central maximum to given order bright line

L = distance from diffraction grating to central maximum

Solution:

Step 1. Find the separation between lines on the diffraction grating. The grating is marked

750 lines per millimeter, so: $d = \left(\dfrac{1\,\text{mm}}{750\,\text{lines}}\right)\left(\dfrac{1\,\text{m}}{1{,}000\,\text{mm}}\right) = 1.33 \times 10^{-6}\,\text{m}$

Step 2. Find the angle.

$$\tan\theta = \frac{x}{L} = \frac{0.54\,\text{m}}{1.00\,\text{m}}$$

$$\theta = 28.4°$$

Step 3. Use the diffraction equation to solve for the laser wavelength.

$$m\lambda = d\sin\theta$$
$$(1)\lambda = (1.33 \times 10^{-6}\,\text{m})(\sin 28.4°)$$
$$\lambda = 6.32 \times 10^{-7}\,\text{m}$$

EXERCISE 14.4

Questions 1–3 use this diagram.

Two small openings on the left are spaced 1 μm apart. Waves coming from the left go through the openings in the same phase and produce an interference pattern with central maximum at point *B* and first order maximum at point *A*.

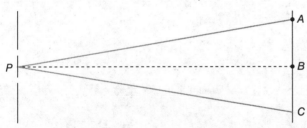

1. Another point where a maximum could be located:

   A. halfway between A and B
   B. point C
   C. halfway between points P and A
   D. halfway between points P and B

2. If the two openings at P are moved farther apart, the position A of the first order maximum:

   A. will remain the same distance from B
   B. will be closer to B
   C. will be farther from B
   D. depends only on the wavelength

3. If a light source with a longer wavelength is used instead:

    A. the distance AB will be less
    B. the distance AB will be greater
    C. the distance AB will be the same

4. Which of the following is <u>not</u> an example of light diffraction?

    A. We see light from the Sun for a short time after sunset.
    B. Compact disks appear to have a rainbow of colors when light hits them.
    C. X-rays are used to determine the spacings between atoms in crystals.
    D. White objects appear to be red when viewed through a red filter.

Atomic and Nuclear Physics

- **Fundamental Forces**
- **Atomic Particles**
- **Radioactive Decay and Half-Life**
- **Absorption and Emission Spectra**
- **Conservation of Energy and Conservation of Charge**
- **Wave-Particle Duality**

This chapter provides only a brief introduction to some topics in modern physics that often appear in physics texts but are usually not considered part of an introductory course. Some common terms are defined as a reference for students who may want to search traditional physics references for more information. A few worked examples help to further explain some advanced topics such as photoelectric effect, emission spectra, Compton effect, and the wave-particle duality.

Fundamental Forces

What physicists refer to as the *fundamental forces in nature* are generally classified as the following four, though they are often recombined into fewer. Other forces, such as the normal force or friction force or tension force, can be explained in terms of the electromagnetic force. The fundamental forces are listed and briefly described here in order from strongest to weakest.

Strong Force: The nuclear force that binds the nucleons (protons and neutrons) together in the nucleus of the atom. This is the strongest of the forces in nature but is short-ranged, affecting only particles within the nucleus.

Weak Force: The nuclear force that is believed to be responsible for beta decay of the nucleus. This force is also effective only within the nucleus.

Electromagnetic Force: As described in Chapter 11, this force is responsible for interactions between particles with charges, such as the attraction of the positive nucleus of an atom to the electrons in the atom outside of the nucleus. It is a field force that extends to infinity.

Gravitational Force: Described in Chapter 4, this is the weakest force of the fundamental forces but probably the most familiar. It is responsible for the attraction of all objects in the universe to each other. It is also a field force that extends to infinity.

1. The weak nuclear force is:

   A. the force that loosely binds outer-shell electrons to the nucleus of an atom
   B. the force that holds the nucleus together
   C. the force that causes protons in the nucleus to repel each other
   D. the nuclear force responsible for beta decay

2. Which of the following fundamental forces is considered to be the strongest (at short range)?

   A. gravitational force
   B. weak nuclear force
   C. strong nuclear force
   D. electromagnetic force

3. Which of the following fundamental forces is actually responsible for friction?

   A. gravitational force
   B. weak nuclear force
   C. strong nuclear force
   D. electromagnetic force

4. Which of the following fundamental forces prevents the protons in the nucleus, which are positive, from repelling each other?

   A. gravitational force
   B. weak nuclear force
   C. strong nuclear force
   D. electromagnetic force

5. Which of the fundamental forces describes the attraction that all objects in the universe have for each other?

   A. gravitational force
   B. weak nuclear force
   C. strong nuclear force
   D. electromagnetic force

Atomic Particles

There are many particles that physicists encounter inside the atom, in nuclear experiments, and throughout the universe. Here are just a few that are mentioned most often in introductory physics.

Proton: Positively charged particle found in the nucleus. Each has over 1,800 times the mass of an electron. They are not considered fundamental particles, since they are made up of quarks.

Neutron: Particle with no net charge found in the nucleus. Neutrons are about the same mass as protons and are also made up of quarks.

Electrons: Fundamental particles with a negative charge. They are found in the atom, held in energy states in the atom by the positive net charge on the nucleus.

Gamma: Electromagnetic radiation often emitted during nuclear processes.

Neutrino: Neutral particle with negligible mass.

Quark: Believed to make up many different particles such as protons, neutrons, and mesons but are always bound to each other, not found isolated in nature.

Alpha Particle: High-energy particle made up of a helium nucleus, i.e., two protons and two neutrons. Often given off by nuclear reactions or nuclear decay.

Beta Particle: High-energy electrons often given off by nuclear reactions or nuclear decay.

Positron: A positively charged particle with the same amount of charge and mass as an electron.

EXERCISE 15.2

1. Which of the following is a fundamental particle?

   A. proton
   B. neutron
   C. electron
   D. deuteron

2. Which is not a true statement about atomic nuclei?

   A. All atomic nucleons are protons and neutrons.
   B. All atomic nuclei have approximately the same density, regardless of composition.
   C. All nuclei of a given element have the same number of protons.
   D. All nuclei of a given element have the same total number of nucleons.

3. The symbol $_2\text{He}^4$ for the element helium means:

   A. An atom of helium has 2 protons and 4 electrons.
   B. A helium nucleus has 2 protons and 4 neutrons.
   C. A helium nucleus has 2 protons and 2 neutrons.
   D. An atom of helium has 2 protons and a total of 4 charged nucleons.

4. A particle that has the same mass as an electron but the opposite charge:

   A. proton
   B. deuteron
   C. positron
   D. quark

5. $_{92}^{238}\text{U}$ is an isotope of uranium that has nuclear composition of:

   A. 92 protons and 92 neutrons and 146 nucleons
   B. 92 protons and 146 neutrons and 146 nucleons
   C. 92 protons and 92 neutrons and 238 nucleons
   D. 92 protons and 146 neutrons and 238 nucleons

Radioactive Decay and Half-Life

Radioactive Decay: The spontaneous breakdown of the nucleus into smaller components by emission of particles.

Half-life: The time it takes for 50% of a certain amount of the nuclei of an element to break down into smaller components.

Isotopes: Forms of the same element that have different numbers of neutrons in their nuclei, so that they have the same atomic number but different mass number. For example, **deuterium** is an isotope of hydrogen that has one proton and one neutron in its nucleus.

EXAMPLE—Radioactive Half-Life

Cs-134, an unstable isotope of cesium, has a half-life of 2 years. How much of a 2-g sample in 2016 would likely still remain in 2022?

Solution:

The time from 2016 to 2022 is 6 years, which would be 3 half-lives. The amount of radioactive Cs-134 would likely be cut in half 3 times—from 2g to 1 g, then from 1 g to 0.5 g, and finally from 0.5 g to 0.25 g.

EXERCISE

15.3

1. The radioactive element 224-Radium has a half-life of 3.66 days. What can you say about one nucleus of radium after 3.66 days?

   A. Half of the nucleus has decayed.
   B. It will only decay when a second nucleus is brought into contact with it.
   C. There is a 50% chance it will have decayed.
   D. It won't decay prior to 3.66 days, but should decay after that time.

2. What nuclear transformation occurs during beta decay?

   A. A proton decays into a neutron and an electron.
   B. An electron and proton combine to form a neutron.
   C. A neutron decays into a proton and an electron.
   D. An electron and a neutron combine to form a proton.

Questions 3–5 refer to the following graph.

The plot below shows the decay curve for a 100-g sample of polonium-210.

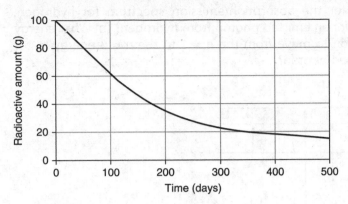

3. Which is the best estimate of the half-life of Po-210 from the plot?

    A. 500 days
    B. 250 days
    C. 180 days
    D. 140 days

4. If an atom of Po-210 ($^{210}_{84}$Po) gives off an alpha ($^4_2$He), predict the product nucleus.

    A. X^{206}_{82}
    B. X^{82}_{206}
    C. X^{214}_{86}
    D. X^{86}_{214}

5. From the plot above, if there is 200 g of polonium-210 at $t = 0$, what is the likely amount of polonium-210 remaining in the sample after 350 days?

    A. 10 g
    B. 20 g
    C. 50 g
    D. 100 g

Absorption and Emission Spectra

Quantized Energy States for Electrons: Electrons within the atom have energy that correlates to an average distance from the nucleus. These energy states are often represented in diagrams showing "energy levels" for the electrons. Electrons can only transition from one energy level to another by absorbing or emitting amounts of energy corresponding to the difference in energy for that transition.

Energy of Photon Emission and Absorption: In order to make energy level transitions, electrons must emit or absorb specific amounts of energy corresponding to the energy state that each level represents. The difference in energy between two levels is $\Delta E = E_2 - E_1$ and the energy difference corresponds to a frequency of light absorbed or emitted: $\Delta E = hf$ or $\Delta E = hc/\lambda$.

Absorption and Emission Spectra: The energy states for atoms of different elements are unique, so the transition energies and spectra are identifiable for specific elements.

EXAMPLE—Electron Energy States

Shown below is the energy diagram for the absorption/emission spectrum for hydrogen. (a) What is the smallest energy possible for an emitted photon from hydrogen? (b) What energy photon must be absorbed for an electron to move from the $n = 1$ to the $n = 3$ energy state? (c) What is the wavelength of the absorbed photon?

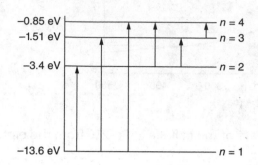

Solution:

(a) Longer arrows represent higher energy transitions, and shorter arrows represent lower energy. For a photon to be emitted, the electron must make the transition from higher energy to lower energy. On this diagram, that transition is from $n = 4$ to $n = 3$: $\Delta E = E_2 - E_1 = -1.51 \text{ eV} - (-0.85 \text{ eV}) = 0.85 - 1.51 = -0.66 \text{ eV}$. The negative sign means the energy is emitted in the form of a photon of that energy.

(b) $\Delta E = E_2 - E_1 = -3.4 - (-13.6) = 13.6 - 3.4 = 10.2 \text{ eV}$. This is the energy of the absorbed photon.

(c) $\Delta E = hc/\lambda$. Since the energy is in electron volts, use the value of hc from Appendix A.

$$\Delta E = \frac{hc}{\lambda} = 10.2 \text{ eV} = \frac{1.24 \times 10^3 \text{ eV} \cdot \text{nm}}{\lambda}$$

$$\lambda = 121.6 \text{ nm}$$

{Note: This is a shorter wavelength than the visible range from about 400–700 nm, so it is in the ultraviolet range.}

1. The energy level diagram represents the first four energy states and possible transitions for an electron in an atom. Which transition would emit a photon with the highest frequency?

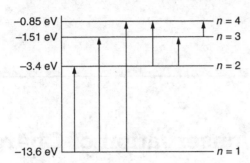

A. from $n = 2$ to $n = 1$
B. from $n = 3$ to $n = 2$
C. from $n = 4$ to $n = 1$
D. from $n = 4$ to $n = 3$

2. Shown below is the energy diagram for the absorption/emission spectrum for hydrogen. What energy transition would emit a photon with the shortest wavelength?

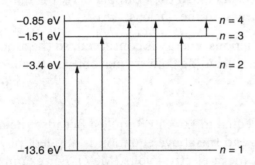

A. from $n = 2$ to $n = 1$
B. from $n = 3$ to $n = 2$
C. from $n = 4$ to $n = 1$
D. from $n = 4$ to $n = 3$

3. Using the energy diagram for the hydrogen atom above, which energy level transition would absorb the photon of highest frequency?

A. from $n = 1$ to $n = 2$
B. from $n = 1$ to $n = 4$
C. from $n = 4$ to $n = 1$
D. from $n = 2$ to $n = 3$

4. An electron changes energy states from higher energy E_2 to lower energy E_1. Which is a correct expression for the wavelength of light emitted in this transition?

A. $\dfrac{hf}{E_2 - E_1}$

B. $\dfrac{hc}{E_2 - E_1}$

C. $\dfrac{f}{E_2 - E_1}$

D. $\dfrac{E_2 - E_1}{hc}$

Conservation of Energy and Conservation of Charge

Fission: Splitting of a heavy, unstable nucleus into two lighter nuclei with the emission of large amounts of energy.

Fusion: Combination of lighter nuclei into a heavier nucleus with the release of huge amounts of energy.

Conservation of Charge: In nuclear reactions, the total charge prior to the reaction is equal to the total charge after the reaction. When balancing a nuclear equation, the charge is usually written on the lower left or right of the symbol for the element. In the example here, the 85 refers to atomic number 85, which means 85 protons (+ charges) in the nucleus: $_{85}At^{218}$

Conservation of Mass-Energy: During nuclear reactions, energy is conserved, so the amount of energy given off or absorbed must be accounted for, including conversions of mass to energy:

$$E = mc^2$$

Electron volt: An electron volt is a unit of energy equal to 1.6×10^{-19} joules. Because the masses of particles used in nuclear reactions are so small and the above equation is applied, physicists often use E/c^2 as a unit of mass, with E in units of mega-electron volts (MeV). For example, the mass of an electron is 9.1×10^{-27} kg, which is the same as 0.511 MeV/c^2.

Conservation of Energy: During processes such as the photoelectric effect or the Compton effect, total energy must be accounted for throughout.

Photoelectric Effect: When struck by certain frequencies of light, metals may release electrons and produce a measurable electric current. Conservation of energy applies, because the energy of the incoming light (hf) must be equal to the work required to release the electrons from the metal (**work function, Φ**) plus the kinetic energy of the electrons producing the current:

$$hf = \Phi + KE$$

As a minimum, the incoming light must have enough energy to release the electrons—even if a current is not produced (i.e., the KE of the electrons is zero). This minimum light energy is the **threshold frequency, f_o,** where $hf_o = \Phi$.

Compton Effect: Collision of a photon with a particle such as an electron conserves energy. If energy is given off in another form such as gamma radiation, then the recoiling photon will have a longer wavelength (i.e., less energy).

Pair Annihilation: Two oppositely charged particles collide in such a way that the charges of the particles cancel and the mass of the particles is converted to energy by the equation:

$$E = mc^2$$

EXAMPLE—Nuclear Equations

$$^6_3\text{Li} + ^2_1\text{H} \rightarrow 2\,^4_2\text{He} + 22.2 \text{ MeV}$$

In the nuclear reaction here, a lithium nucleus with atomic number 3 and mass number 6 combines with a hydrogen nucleus (deuterium) with atomic number 1 and mass number 2 to form two helium nuclei (alpha particle) with atomic number 2 and mass number 4 along with a release of 22.2 MeV of energy. (An electron volt is a measure of energy equal to 1.6×10^{-19} J.)

The atomic numbers on the lower part of the symbols represent protons in the nuclei that are positively charged. The total charge of +4 on the left side equals the total charge of +4 on the right side. The upper symbols are mass numbers (but not actual masses) of the nuclei. Notice that the total mass number on left and right are equal.

However, the actual masses of the two helium nuclei on the right are slightly less than the total mass on the left. That "missing mass" has actually been converted to large amounts of energy, by the equation $E = mc^2$.

EXAMPLE—Photoelectric Effect

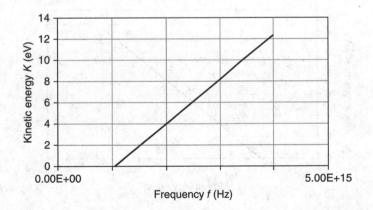

Use the graph of kinetic energy as a function of frequency for light shining on a metal to determine the work function of this metal.

Solution:

$$hf = \Phi + KE$$

The threshold frequency is the frequency where the kinetic energy is zero. In this case, that is a frequency of about 1×10^{15} Hz. Therefore, the energy of the photons striking the metal (hf) is equal to the work function. The work function, Φ, for this metal is 6.63×10^{-19} J, since $hf = \Phi$. (This work function might be given as 4.1 electron volts.)

1. As a result of the emission of an alpha particle, an atomic nucleus:

   A. increases its atomic number by 2 and increases its mass number by 4
   B. decreases its atomic number by 2 and decreases its mass number by 4
   C. increases its atomic number by 2 and decreases its mass number by 4
   D. decreases its atomic number by 2 and increases its mass number by 4

2. As a result of the emission of a beta particle, an atomic nucleus:

   A. increases its atomic number by 2 and increases its mass number by 1
   B. increases its atomic number by 1 but doesn't change its mass number
   C. increases its atomic number by 1 and decreases its mass number by 1
   D. decreases its atomic number by 1 and increases its mass number by 1

3. During radioactive emission of gamma radiation in which no other particles are emitted, the gamma photons are emitted in pairs. Which conservation law would be defied if only one gamma photon were emitted?

   A. conservation of mass
   B. conservation of charge
   C. conservation of linear momentum
   D. conservation of energy

4. Below is a chart representing data from a photoelectric effect experiment. Estimate the minimum (or cutoff) frequency of light that will cause ejection of electrons from zinc.

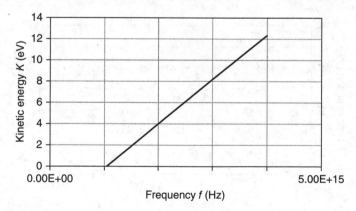

   A. 5.0×10^{14} Hz
   B. 2.5×10^{15} Hz
   C. 2.0×10^{15} Hz
   D. 1.0×10^{15} Hz

5. One step in the radioactive decay series of uranium-235 to lead-207 is the decay of bismuth-215. In this step, $_{83}Bi^{215}$ gives off an alpha particle and a beta particle to produce:

   A. $_{85}At^{218}$
   B. $_{82}Pb^{211}$
   C. $_{83}Bi^{211}$
   D. $_{84}Po^{211}$

6. A radioactive sample is placed in a cloud chamber so that the paths of radioactive decay products can be seen as "tracks" in the chamber. A magnetic field is set up across the cloud chamber to help identify the decay products. Which of the following radioactive decay products would never change direction in the magnetic field?

A. alpha particles
B. beta particles
C. gamma rays
D. positrons

7. In the nuclear fission reaction below, a uranium atom is bombarded with a neutron to produce an atom of barium and an atom of krypton. What is the other product produced?

$$^{235}_{92}U + ^{1}_{0}n \rightarrow ^{139}_{56}Ba + ^{94}_{36}Kr + \underline{\hspace{1cm}}$$

A. an electron
B. two electrons
C. a proton
D. three neutrons

8. In a nuclear fusion reaction, four protons (hydrogen nuclei) combine to form a helium nucleus, two neutrinos, and another product. Which of the following could be missing product in this reaction?

$$4^{1}_{1}H \rightarrow ^{4}_{2}He + 2^{0}_{0}\eta + \underline{\hspace{1cm}}$$

A. helium nucleus
B. two protons
C. two positrons (positive electrons)
D. two beta particles (negative electrons)

9. According to some scientists, the fusion of deuterium with tritium to create helium-4 may be a source of energy for the future, since the reaction gives off large amounts of energy. What is the other product of this fusion reaction?

$$_{1}H^{2} + _{1}H^{3} \rightarrow _{2}He^{4} + \underline{\hspace{1cm}?\hspace{1cm}} + energy$$

A. a proton
B. a neutron
C. an electron
D. a positron

10. A photon of light collides with a stationary electron. Using conservation of linear momentum and conservation of mass-energy principles, what is the expected result?

   A. The electron recoils with kinetic energy K and a photon with a longer wavelength is produced.

   B. The electron recoils with kinetic energy K and the photon with a shorter wavelength is produced.

   C. The electron recoils with kinetic energy K and the photon retains its previous speed and wavelength after the collision.

   D. The electron remains stationary and the photon rebounds from it with a shorter wavelength.

11. The number 931 MeV/c^2 is a measurement of a particle's:

   A. momentum
   B. mass
   C. energy
   D. velocity

Wave-Particle Duality

Photon: A quantum of light that has both particle and wave behavior. A photon carries energy proportional to its frequency: $E = hf$

Momentum of a Photon: We are familiar with the momentum of objects moving at speeds in the laboratory: $p = mv$. By the nature of wave-particle duality, waves also exhibit properties of particles. The momentum of a particle moving near the speed of light is also a mass times a velocity—or a "mass equivalent" times the velocity of light:

$$p = \left(\frac{E}{c^2}\right)(c) = \frac{E}{c} = \frac{hf}{c} = \frac{h}{\lambda}$$

Wave-Particle Duality: Particles can have properties of waves, as demonstrated by the Davisson-Germer experiment, where fast-moving electrons were shown to diffract through a small opening, just as waves diffract through small openings.

DeBroglie Wavelength: Moving particles can have a wavelength, called the deBroglie wavelength, that is:

$$p = \frac{hf}{c} = \frac{h}{\lambda}$$

1. Which of the following is a correct expression for the momentum of a photon?

   A. $p = \dfrac{h}{f}$

   B. $p = \dfrac{h}{c}$

   C. $p = hf$

   D. $p = \dfrac{h}{\lambda}$

2. Calculate the deBroglie wavelength of an electron moving at 1×10^6 m/s.

   A. 7.28×10^{-10} m
   B. 4.1×10^{-9} m
   C. 728 m
   D. 4.14×10^{-21} m

3. Calculate the energy of a photon with wavelength 700 nm.

   A. 9.5×10^{-19} J
   B. 9.5×10^{-28} J
   C. 2.8×10^{-19} J
   D. 2.8×10^{-28} J

Fundamental Constants and Useful Information

Universal gravitational constant: $G = 6.67 \times 10^{-11}$ N-m^2/kg^2

Coulomb's constant: $k = \dfrac{1}{4\pi\varepsilon_o} = 9.0 \times 10^{-9}$ N-m^2/C^2

Gas law constant: $R = 8.31$ J/mol-K

Atmospheric pressure (sea level): 1.0×10^5 Pa

Electric permittivity in vacuum: $\varepsilon_o = 8.85 \times 10^{-12}$ C^2/N-m^2

Magnetic permeability in vacuum: $\mu_o = 4\pi \times 10^{-7}$ T-m/A

Planck's constant: $h = 6.63 \times 10^{-34}$ J·s $= 4.14 \times 10^{-15}$ eV-nm

$\qquad\qquad hc = 1.24 \times 10^3$ eV-nm $= 1.99 \times 10^{-25}$ J-m

Speed of light in vacuum: $c = 3.0 \times 10^8$ m/s

Boltzmann's constant: $k_B = 1.38 \times 10^{-23}$ J/K

Mass of electron: 9.11×10^{-31} kg

Mass of proton = mass of neutron: 1.67×10^{-27} kg

Electron charge: -1.6×10^{-19} C

Mass of Earth: 5.97×10^{24} kg

Mass of Sun: 2×10^{30} kg

Mass of Moon: 7.35×10^{22} kg

Earth-Moon distance (average): 3.84×10^8 m

Earth-Sun distance (average): 1.49×10^9 m

Earth radius (average): 6.37×10^6 m

Gravitational field on Earth's surface: 9.81 N/kg or 9.81 m/s^2

Metric Units and Conversions

Common Metric Prefixes

pico-	(p)	10^{-12}
nano-	(n)	10^{-9}
micro-	(μ)	10^{-6}
milli-	(m)	10^{-3}
centi-	(c)	10^{-2}
deci-	(d)	10^{-1}
deka-	(dk or D)	10^{1}
hecto-	(H)	10^{2}
kilo-	(k)	10^{3}
mega-	(M)	10^{6}
giga-	(G)	10^{9}

Useful Information

1 inch = 2.54 cm

3.28 ft = 1 meter

1 mile = 1609 m

1.06 qt = 1 liter

1 year = 31,536,000 seconds

1 day = 86,400 seconds

1 cubic meter = 1,000 liters

Answers to Chapter Exercises

Note: A value of 9.8 m/s^2 for g is used in most solutions, but 10 m/s^2 will lead to a correct answer choice.

Chapter 1

Exercise 1.1

1. B   sin A = 21/29 Opposite side divided by hypotenuse
2. B   Use Pythagorean Theorem $(AB)^2 + 21^2 = 29^2$ AB = 20
3. A   Use Pythagorean Theorem $8^2 + 6^2 = d^2$ $d = 10$
4. D   tan angle = opposite/adjacent = height of pole/ground distance = 8/6
       angle = inverse tan (8/6)   angle = 53°
5. D   You should identify a larger triangle and a smaller triangle that have different sides but the same angles. Use these similar triangles to set up ratios. $6/2 = (3 + x)/x$
       Cross multiply and solve. $6x = 6 + 2x$ $x = 3/2$

Exercise 1.2

1. D $6x + 6 = 14x - 6$ $12 = 8x$ $x = 12/8 = 1.5$
2. C $x^2 = 6x$ Factor method: $x(x - 6) = 0$ $x = 0$ or 6
3. D $2x^2 - x = 0$ Factor: $x(2x - 1) = 0$ $x = 0$ or ½
4. A $2t^2 - 6t = 0$ Factor: $2t(t - 3) = 0$ $t = 0$ or 3

5. B $F = \dfrac{m\mathbf{v}^2}{r}$ $r = \dfrac{m\mathbf{v}^2}{F}$

6. D $1 \, \cancel{cm} \left(\dfrac{1 \, \cancel{m}}{100 \, \cancel{cm}} \right) \left(\dfrac{10^6 \, \mu m}{1 \, \cancel{m}} \right) = 10^4 \, \mu m$

7. A $\left(\dfrac{60 \, \cancel{ft}}{1 \, s} \right) \left(\dfrac{1 \, m}{3.28 \, \cancel{ft}} \right) = 18.29$ m/s

8. C   This is the incorrect statement because $m \neq \left(\dfrac{m}{s^2} \right)(s^3)$ when you cancel seconds.

9. A   Kilometers per hour would be used. For example, if you convert 75 mi/h to km/h, it is about 121 km/h.

241

10. B $\quad x = vt = (186,000 \text{ mi/sec})(31,536,000 \text{ sec/year}) = 5.87 \times 10^{12}$ miles per year $= 1$ light year

11. C Add the numbers together and then divide by 4 to get the average. Round to the 0.1 place, since that is the least rounded value on the numbers. Ans: 2.4 m

12. C Percent error is the difference between the accepted and experimental values divided by the accepted value.

$$\frac{9.81 \text{ m/s}^2 - 9.76 \text{ m/s}^2}{9.8 \text{ m/s}^2} \times 100\% = 0.5\%$$

13. A Add the numbers to get 20.086. Then round to the 0.01 place, since that is the least rounded value on any of the numbers.

14. A The square root of 4 is 2, so squaring 2 obtains 4. When multiplying, the number of digits in the factor with the least number of digits decides the number of digits in the product. So you would multiply 2.0 times 2.0 to get 4.0 with the correct number of digits.

15. B $\quad K = \frac{1}{2}mv^2 \qquad 2K = mv^2 \qquad \dfrac{2K}{m} = v^2 \qquad v = \sqrt{\dfrac{2K}{m}}$

16. D $\quad \left(\dfrac{35 \text{ mi}}{1 \text{ h}}\right)\left(\dfrac{1609 \text{ m}}{1 \text{ mi}}\right)\left(\dfrac{1 \text{ h}}{3600 \text{ s}}\right) = 15.6$ m/s

17. D $\quad 150 \text{ cm} = 1.50 \text{ m}$, since 1 m $= 100$ cm

18. D $\quad (1 \times 10^6 \text{ mm})\left(\dfrac{1 \text{ m}}{1,000 \text{ m}}\right) = 1,000$ m

Exercise 1.3

1. B The curve is of the form $y = k\sqrt{x}$, so that is what should be plotted: y-values on the y-axis and take square root of all x-values to replot on the x-axis.

2. D The curve is of the form $y = kx^2$, so replot the y-values on the y-axis and square all the x-values to plot on the x-axis.

3. A Use values from the graph, e.g., if $x = 10$, then $y = 200$, and $y = 2x^2$

4. C When $x = 10$, $y = 200$ m.

5. B II is of the form $y = kx^2$.

6. A Slope is rise/run, which is N/m. Area is length times width, so that would be N-m.

Chapter 2

Exercise 2.1

1. C Add Vector **A** to the opposite of Vector **B** (just the opposite direction)—but still add tip to tail. Then draw the resultant from the tail of Vector **A** to the tip of **B**.

2. A Add Vector **A** to Vector **B** tip to tail and draw the resultant from the tail of **A** to the tip of **B**. These can be added in either order and still get the same resultant.

3. C The two vectors are perpendicular, so use the Pythagorean Theorem. The question asks for speed, so the direction is not needed on the answer.

4. C The two vectors are perpendicular, so use the Pythagorean Theorem. The answer is just slightly over 100, and the wind is toward the east, so the resultant is slightly east of north.

5. B To add opposite vectors in one dimension, just add positive and negative values. Displacement here is the sum of −100 m and +32 m, which is −64 m.

Exercise 2.2

1. A The best answer is A. The object might go 10 m forward and then go 10 m back. If there is no displacement then the average velocity is zero, but the object has moved a distance of 20 m.

2. A Displacement is final position minus original position. Since the final and initial positions are the same, the displacement is zero.

3. A Displacement is zero, so average velocity is zero.

4. B x(final) − x(original) = −3 − 3 = −6 m

5. A v(average) = [v(original) + v(final)]/2 v(average) = (1 + −1)/2 = 0

Exercise 2.3

1. D $\mathbf{x}_f = \mathbf{x}_o + \mathbf{v}_o t + \frac{1}{2}\mathbf{a}t^2$

$2 = 0 + 0 + \frac{1}{2}\mathbf{a}(4)^2$

$\mathbf{a} = \frac{1}{4}$ m/s^2

$\mathbf{v}_f = \mathbf{v}_o + \mathbf{a}t$

$\mathbf{v}_f = 0 + \frac{1}{4}(4)^2 = 4$ m/s

$\Delta\mathbf{x} = (4)(4) + \frac{1}{2}(\frac{1}{4})(4)^2 = 18$ m

2. C (Definition)

3. D For example, if the line is horizontal, the $\mathbf{v} = 0$ and $\mathbf{a} = 0$.

4. A Gravity and g are downward throughout, but vertical velocity is positive on the way up, zero at the top, and negative on the way down.

5. B. The gravitational force is always the same—downward—regardless of the motion.

6. A Use the kinematics equation:

$\Delta\mathbf{y} = \mathbf{v}_{oy}t + \frac{1}{2}\mathbf{a}t^2$

-500 m $= 0 + \frac{1}{2}(-9.8$ m/s$^2)t^2$

$t = 10$ s

7. B Use the equation:

$\Delta\mathbf{y} = \mathbf{v}_{oy}t + \frac{1}{2}\mathbf{a}t^2$

$0 = \mathbf{v}_{oy}(4$ s$) + \frac{1}{2}(-9.8$ m/s$^2)(4$ s$)^2$

$\mathbf{v}_{oy} = 19.6$ m/s

Exercise 2.4

1. C The tangent to the curve is acceleration. The curve upward shows increasing slope, which is increasing acceleration, and the line itself shows increasing values of speed.

2. D Velocity value is zero at $t = 0$ and $t = 7$ s.

3. A Negative velocity values would be below the x-axis, from 7–9 s.

4. C Area between the graph line and the x-axis is the distance traveled or change in position. For the first 3 seconds, that area is a triangle with base 3 s and height 8 m. Distance = area = ½ (3 s)(8 m) = 12 m

5. A Acceleration is the slope of a velocity versus time graph.
   Slope = rise/run = 8 m/s/3 s = 2.67 m/s².

6. D Displacement = x(final) – x(initial) = 3.3 m – 0.9 m = 2.4 m

7. A On a position versus time graph, slope is velocity, so the steepest slope would be the greatest velocity

8. C The points are A, D, and E. Look for points on the graph where $v = 0$.

9. C The velocity is zero at A and then goes from positive to zero at D and then from zero to negative at E.

10. D A greater rate of change in position would be a greater velocity, so steeper slope means a greater velocity.

11. C $\Delta x = vt = (-2$ m/s$)(4$ s$) = -8$ m

12. B Displacement = x(final) – x(initial) = 10 = 0 = 10 m. Path in between doesn't matter.

13. B Slope of a Displacement versus time graph is velocity. (The units of rise/run would be m/s.)

14. B The line has a constant slope, which is the velocity.
   Velocity = rise/run = −5 m/5 s = −1 m/s

15. D Both the speed and direction stay constant, so velocity is constant in this case.

16. D The slope of the curve at that point can be approximated as somewhat linear, so slope is about −3.5/0.5 or −7 m/s.

17. B On a Position versus time graph, the direction of movement changes when the line or curve changes the sign of slope.

Chapter 3

Exercise 3.1

1. B Newton's third law: action and reaction. The reaction force here is equal in magnitude and opposite in direction.

2. D You angle your foot slightly and push backward and downward, so the reaction forces that propel you forward (Newton's third law) are friction force and a component of normal force from the floor.

3. D The forces on box M are: normal force from box m downward, normal force from floor upward, gravitational force on M, and friction force from box m that makes it move to the right. (Note: Box m pushes down on M, but the weight of box m is defined as the Earths force on the box. The gravitational forces between the boxes are not significant compared to the gravitational force of Earth on the boxes.)

4. C The forces on box m are: force $\mathbf{F}$ to the right, normal force upward from box M, friction force from box M to the left (since m is pulled to the right), gravitational force of Earth on box m downward.

5. C The action force is the gravitational force of Earth on M, and reaction force is the gravitational force of M on the Earth.

Exercise 3.2

1. A Make east positive. The net force is 10.5 N east – 6.4 N west, or 4.1 N east. Acceleration is net force divided by mass or 4.1 N/3 kg = 1.37 N east.

2. B $\mathbf{F} = 6$ N when $\mathbf{a} = 10$ m/s^2, so $m = \mathbf{F}/\mathbf{a} = 6/10 = 0.6$ kg.

3. C Newton's third law: regardless of the motion of the system, the internal forces are equal in magnitude and opposite in direction. The car exerts the same force on the trailer as the trailer exerts on the car.

4. D Mass is 2 kg and $\mathbf{a} = 5$ m/s^2, so net force equals $m\mathbf{a} = 10$ N.

   A force forward of 12 N and backward (friction) of 2 N provides the necessary 10 N.

   In other words, for a 2-kg object to accelerate at 5 m/s/s, the *net* force must be 10 N forward.

5. C The two forces directed up the ramp must equal the only force directed down the ramp, which is a component of the gravitational force.

6. A First find the mass of the elevator. Weight downward is 10,000 N. Mass of elevator = weight/g = 10,000/9.8 = 1,020 kg.

   Net force on elevator = $\Sigma\mathbf{F} = m\mathbf{a}$

   $T - 10,000$ N $= m\mathbf{a}$

   $T = m\mathbf{a} + 10,000$ N $= (1,020$ kg$)(3$ m/s$^2) + 10,000$ N $= 13,060$ N

   The net force on the elevator must be enough to hold the weight and accelerate it upward.

7. D $\mathbf{a} = \dfrac{\mathbf{F}}{m} = \dfrac{10\text{ N}}{15\text{ kg}} = 0.67$ m/s^2

8. D First, find the acceleration and then use that to find the change in velocity. Finally, look for the answer that has that change in velocity.

   $$\mathbf{a} = \frac{\mathbf{F}}{m} = \frac{10\text{ N}}{2\text{ kg}} = 5\text{ m/s}^2 \qquad \Delta\mathbf{v} = \mathbf{a}t = (5\text{ m/s}^2)(5\text{ s}) = 25\text{ m/s}^2\text{m/s}$$

9. D Since the forces are perpendicular, find the net force using the Pythagorean Theorem:

   $$\mathbf{F}^2 = 30^2 + 40^2$$

   $$\mathbf{F} = 50\text{ N}$$

   $$\mathbf{a} = \frac{\Sigma\mathbf{F}}{m} = \frac{50\text{ N}}{35\text{ kg}} = 1.4\text{ m/s}^2$$

10. B Use the system approach with the total mass of the boxes and the total force applied. To pull the system to find the acceleration. They all accelerate the same.

    $$\mathbf{a}_{\text{system}} = \frac{\mathbf{F}_{\text{on system}}}{m_{\text{system}}} = \frac{12\text{ N}}{6\text{ kg}} = 2\text{ m/s}^2$$

    Then apply Newton's second law to the 1-kg block alone: $\mathbf{F} = m\mathbf{a} = (1$ kg$)(2$ m/s/s$) = 2$ N

11. B When accelerating upward, the person needs more force upward than downward to create a net force upward.

Therefore, the scale exerts a force upward that is more than the person's actual weight, and the scale reads more than the actual weight.

12. A $F = ma$ To maintain a constant F in the second case when mass is three times as much, the force must also be three times as much.

13. A $F = ma$ When $F = 100$ newtons and mass is 20 kg, acceleration should be 5 m/s^2.

If the acceleration is only 4 m/s^2, the net force forward must only be 80 N. This means that there is a 20 N force in the negative direction, which is likely the friction force.

14. C The net force in the east-west direction is $+10 + 20 - 30 = 0$. So the only net force is north. Therefore, the acceleration is north: $\mathbf{a} = \mathbf{F}/m = 30$ N/10 kg $= 3$ m/s^2 north.

Exercise 3.3

1. C The net force is zero and acceleration is zero, so a 100-N force forward must be balanced by a 100-N force backward (friction).

2. D The net force is zero and acceleration is zero, so a 12-N force forward must be balanced by a 12-N force backward (friction).

3. A The net force on the box is zero, so force upward on the box equals force downward. The gravitational force on the box causes it to exert a normal force of 500 N downward on the table. By Newton's third law, the table exerts a reaction force of 500 N upward on the box.

4. A Only a component of each cord is upward to support the weight, so each component supports half of the weight. Each component is less that the tension in the cord, so the tension is greater than 50 N.

5. B The only two forces in the horizontal direction must add to zero. The piano is in equilibrium, so the force to the right must be equal to and in opposite direction of the force to the left.

Chapter 4

Exercise 4.1

1. D $\mathbf{F}_{G(Earth)} = \dfrac{GM_E m}{(R_E)^2}$ $\mathbf{F}_{G(Planet)} = \dfrac{G(300{,}000\, M_E)m}{(100\, R_E)^2} = 30\, \mathbf{F}_{G(Earth)}$

2. D The net force on B is least, since one of the other objects exerts a force to the right and the other object exerts a force to the left. For the objects on the ends, the distances and masses are the same for the other two, so the net force on each of the end objects is the same.

3. D The equation for gravitational force is the same, except R in orbit is now $4R$. When that 4 in the denominator is squared, it makes the new value for the force 1/16 as much.

4. A Newton's third law: The forces are equal and opposite.

5. C Newton's third law: The forces are equal and opposite.

Exercise 4.2

1. A $\mathbf{F}_{G(planet)} = mg_{planet} = \dfrac{GM_{planet}m}{R_{planet}^2}$ therefore: $g_{planet} = \dfrac{GM_{planet}}{R_{planet}^2}$

2. C Doubling the radius, R, which is squared, in the equation for gravitation force would make the force ¼ as much. This is an application of the "inverse square law." When the distance is doubled, the force is ¼. (When the distance is tripled, the force is $\frac{1}{9}$, and so on.)

3. A Weight changes in different gravitational fields, but mass does not change.

4. B The effect of rotation on g is to decrease its value. At the geographic North Pole, which is on the rotational axis, there is no rotational effect on g.

5. C All of the other answers depend on g, but a spring performs the same regardless of gravitational field.

Exercise 4.3

1. B $\quad \mathbf{a}_1 = \dfrac{\mathbf{v}^2}{R} \qquad \mathbf{a}_2 = \dfrac{(2\mathbf{v})^2}{2R} = 2\mathbf{a}_1$

2. D $\quad g_{\text{Earth}} = \dfrac{GM_E}{R_E^2} \qquad g_{\text{Planet}} = \dfrac{G(\frac{1}{2}M_E)}{(2R_E)^2} = \frac{1}{8}M_E$

3. D Moving in a uniform circles has constant centripetal acceleration. B and C could also be true.

4. B A net force toward the center of a circle or arc is required for motion in a circular path.

5. B A component of the normal force on a banked curve adds to the friction force—both toward the center of the circular path.

6. C Centripetal force from the gravitational force is toward the center.

7. C The forces they exert on each other are equal in magnitude and opposite in direction.

8. B $\quad \mathbf{a} = \dfrac{m\mathbf{v}^2}{R} = \dfrac{(2{,}000 \text{ kg})(20 \text{ m/s})^2}{100 \text{ m}} = 8{,}000 \text{ m/s}^2$

Exercise 4.4

1. A Inertia keeps the ball moving horizontally at constant speed if we assume there is no net force such as air friction horizontally.

2. D If there is negligible air friction, no force is exerted in the x-direction.

3. A Displacement equals y(final) minus y(initial) and they are both at the ground, so displacement is zero.

4. A The horizontal velocity continues at the peak of the parabola, even though the vertical motion comes to a momentary stop at the object turns to fall back down.

5. C This is the definition of motion that creates a parabola—net force in one dimension and no force in the second dimension. This is true of projectiles or for motion of an object on a flat surface.

6. C $\mathbf{v}$ is constant horizontally at 2 m/s. Use the height to determine time in the air. Remember that time in the air is the same for both the horizontal and vertical motions.

Vertical: $\qquad \Delta\mathbf{y} = \mathbf{v}_{oy}t + \frac{1}{2}\mathbf{a}t^2$

$\qquad\qquad -1 \text{ m} = 0 + \frac{1}{2}(-9.8 \text{ m/s}^2)t^2$

$\qquad\qquad -1 = -4.9\,t^2 \qquad t = 0.45 \text{ s}$

Horizontal: $\qquad \mathbf{X} = \mathbf{v}_{ox}t = (2 \text{ m/s})(0.45 \text{ s}) = 0.9 \text{ m}$

7. C Regardless of horizontal motion, the time for the two objects to fall is the same (only depends on height).

8. A There is no acceleration horizontally, and the acceleration vertically is g. (Whether or not g is negative depends on the sign convention used to work the problem.)

9. C Horizontal velocity stay constant at 2 m/s, but vertical velocity starts at zero and increases as the object falls, depending on height.

$$\mathbf{v}_{fy}^2 = \mathbf{v}_{oy}^2 + 2\mathbf{a}\Delta\mathbf{y}$$
$$\mathbf{v}_{fy}^2 = 0 + (2)(9.8 \text{ m/s}^2)(5 \text{ m})$$
$$\mathbf{v}_{fy} = 9.9 \text{ m/s}$$

10. D Acceleration is g downward at all points during the motion, leaving only B or D as possible answers. Maximum velocity occurs at launch (A) or at landing (E).

Chapter 5

Exercise 5.1

1. D Change in potential energy is equal to gain in kinetic energy:

$$\cancel{m}gh = \tfrac{1}{2}\cancel{m}v^2$$

$$(9.8 \text{ m/s}^2)(0.2 \text{ m}) = \tfrac{1}{2}v^2 \qquad v = 1.98 \text{ m/s}$$

2. B Potential energy plus kinetic energy at the top all becomes kinetic energy at the bottom:

$$PE_{top} + KE_{top} = KE_{bottom}$$
$$mgh + \tfrac{1}{2}mv^2 = KE_{bottom}$$
$$(0.5 \text{ kg})(9.8 \text{ m/s}^2)(20 \text{ m}) + \tfrac{1}{2}(0.5 \text{ kg})(5 \text{ m/s})^2 = 104 \text{ J}$$

A common mistake is to try to cancel mass—but it's not on both sides, so can't cancel.

3. D $\Delta KE = KE_{final} - KE_{original} = \tfrac{1}{2}(2{,}000 \text{ kg})(20 \text{ m/s})^2 - \tfrac{1}{2}(2{,}000 \text{ kg})(10 \text{ m/s})^2 = 300{,}000 \text{ J}$

4. D Show that KE is proportional to square of time, so the plot would be a "power curve."

$$KE = \tfrac{1}{2}mv^2 \qquad \text{and} \qquad \mathbf{v}_f = \mathbf{v}_o + \mathbf{a}t \qquad \text{with} \qquad \mathbf{v}_o = 0$$

Substitute:

$$KE = \tfrac{1}{2}m(\mathbf{a}t)^2$$

5. D Kinetic energy is the sum of the original KE and PE:

$$PE + KE = KE_{final}$$

$$mgh + \tfrac{1}{2}mv^2 = (2 \text{ kg})(9.8 \text{ m/s}^2)(4 \text{ m}) + \tfrac{1}{2}(2 \text{ kg})(3 \text{ m/s})^2 = 87.4 \text{ J}$$

6. D Kinetic energy at the bottom is the sum of the KE and PE at the top.

7. D $U_1 = \dfrac{GMm}{R} \qquad U_2 = \dfrac{G(2\,M)(2\,m)}{(\tfrac{1}{2}R)} = 8\,U_1$

8. C Apply conservation of energy:

$$PE_1 = \tfrac{1}{2}kx^2 = \tfrac{1}{2}(200 \text{ N/m})(0.01 \text{ m})^2 = 0.01 \text{ J}$$
$$PE_2 = \tfrac{1}{2}kx^2 = \tfrac{1}{2}(200 \text{ N/m})(0.02 \text{ m})^2 = 0.04 \text{ J}$$

Easier method: Doubling x, which is squared, will make it four times the energy.

9. A Spring potential energy is changed to kinetic energy:

$$\tfrac{1}{2}kx^2 = \tfrac{1}{2}m\mathbf{v}^2$$

$$k(0.05\text{ m})^2 = (0.1\text{ kg})(2\text{ m/s})^2 \qquad k = 160\text{ N/m}$$

Exercise 5.2

1. A Force and displacement are in the same direction, so force is maximum on this choice.

2. D No net work is done in choices A, B, or C.

3. D The centripetal force is perpendicular to the direction of velocity at every point, so it does no work.

4. B Friction force opposes motion; friction force and displacement are in opposite directions, making work done by friction negative. Work is force times distance, so μmgd has units of force and distance, with μ having no units.

5. A Force and distance are in the same direction, so work is positive and maximum.

6. A Potential energy of the spring minus thermal energy lost by work done against friction equals the remaining kinetic energy:

$$PE - W_f = KE$$

$$\tfrac{1}{2}kx^2 - \mu mgd = \tfrac{1}{2}mv^2$$

$$\tfrac{1}{2}(200\text{ N/m})(0.05)^2 - (0.02)(0.5\text{ kg})(9.8\text{ m/s}^2)(0.05\text{ m}) = \tfrac{1}{2}(0.5\text{ kg})v^2$$

$$0.25\text{ J} - 0.0049\text{ J} = 0.25\, v^2$$

$$v = 0.99\text{ m/s}$$

7. C The potential energy in the spring becomes gravitational potential energy:

$$PE_S = PE_G$$

$$\tfrac{1}{2}kx^2 = mgh$$

$$\tfrac{1}{2}(k)(0.01)^2 = (0.02\text{ kg})(9.8\text{ m/s}^2)(0.08\text{ m})$$

$$k = 313.6\text{ N/m}$$

8. D Work is equal to the change in potential energy plus kinetic energy.

9. B Work equals change in energy. Energy change in A is zero; energy change in B is mgh or 392 J; change in energy in C is zero; and energy change in D is ½ mv^2 or 250 J.

10. D Change in potential energy equals change in kinetic energy, and change in kinetic energy equals work.

$$\Delta PE = \Delta KE = \mathbf{F}d$$

$$mgh = \mathbf{F}d$$

$$(5\text{ kg})(9.8\text{ m/s}^2)(1\text{ m}) = \mathbf{F}(0.01\text{ m})$$

$$\mathbf{F} = 4{,}900\text{ N}$$

Exercise 5.3

1. B Work is negative, since it reduces the KE of the system.

2. C Friction force does negative work, since it reduces the kinetic energy and converts it to thermal energy.

3. C Friction opposes the ball's motion and does negative work, decreasing the ball's kinetic energy.

4. C Gravity decreases potential energy, so $W = mgh = (2\text{ kg})(9.8\text{ m/s}^2)(10\text{ m}) = 196\text{ J}$.

5. D No work is required to keep the satellite moving, since it has inertia. The centripetal force keeps the satellite moving in a circle, but the centripetal force does no work.

Exercise 5.4

1. A From the graph, the maximum PE is about 4900 J. Set that equal to mgh and find m.

   $4900 \text{ J} = mgh = (m)(9.8 \text{ m/s}^2)(490 \text{ m})$ $m = 1 \text{ kg}$

2. A From the graph, PE = KE at $t = 7$ s, so each has ½ of the total energy.

3. D Kinetic energy at 2 s is about 200 J, and KE at 4 s is closer to 800 J.

4. B Gravitational potential energy is converted to kinetic energy.

   $\Delta mgh = \Delta \frac{1}{2}mv^2$

   $(2 \text{ kg})(9.8 \text{ m/s}^2)(4 \text{ m}) = 78.4 \text{ J}$ kinetic energy

   The object lands with 78.4 J of kinetic energy but leaves the floor with 10% less, or 7.84 J

   Less leaving the floor, the object has 70.6 J of KE.

5. B Using the method on the previous problem, the ball leaves the floor with 70.6 J of KE.

   Convert that back to PE after the bounce: $70.6 \text{ J} = mgh = (2 \text{ kg})(9.8 \text{ m/s/s})(h)$, $h = 3.6 \text{ m}$

Exercise 5.5

1. A Power equals work divided by time or energy divided by time or average force times average velocity. In A, power is $mgh/2 = 49$ W; power in B is equal to Fv or 30 W. Power in choice C is $\Delta KE/t$ or 2 W, and power in D is just a steady 10 W.

2. C Work and power will have the same sign, since $P = W/t$.

3. A $W = \Delta KE = \frac{1}{2}mv^2 = \frac{1}{2}(2{,}200 \text{ kg})(30 \text{ m/s})^2 = 990{,}000 \text{ J}$

4. C $P = W/t = 990{,}000 \text{ J}/40 \text{ s} = 24{,}750 \text{ W}$

5. A Power equals change in kinetic energy divided by time.

$$P = \frac{\Delta KE}{t} = \frac{KE_f - KE_o}{t} = \frac{\frac{1}{2}mv_f^2 - \frac{1}{2}mv_o^2}{t} = \frac{\frac{1}{2}(2 \text{ kg})(3^2 - 2^2)}{10 \text{ s}} = 0.5 \text{ W}$$

Chapter 6

Exercise 6.1

1. D Apply the center of mass formula, as described in the first section of Chapter 6:

$$x = \frac{(2)(0) + (6)(5) + (2)(2)}{10} = 3.4$$

$$y = \frac{(2)(7) + (6)(6) + (2)(2)}{10} = 5.4$$

2. B Apply the changes to momentum and to kinetic energy:

   $\mathbf{p}_o = m\mathbf{v}$ $KE_o = \frac{1}{2}mv^2$

   $\mathbf{p}_{new} = (2\ m)(2\ \mathbf{v}) = 4\ m\mathbf{v} = 4\ \mathbf{p}_o$ $KE_{new} = \frac{1}{2}(2\ m)(2\ \mathbf{v})^2 = 8(\frac{1}{2}mv^2) = 8\ KE_o$

3. B Convert, using known values from Appendix B and cancelling units:

$$\mathbf{v} = (60 \text{ mi}/\cancel{h})\left(\frac{1 \cancel{h}}{3{,}600 \text{ s}}\right)\left(\frac{1{,}609 \text{ m}}{1 \cancel{mi}}\right) = 26.8 \text{ m/s}$$

$$\mathbf{p} = m\mathbf{v} = (2{,}000 \text{ kg})(26.8 \text{ m/s}) = 53{,}633 \text{ kg} \cdot \text{m/s}$$

4. C Momentum is a vector but kinetic energy is a scalar quantity.

5. B Just by examining the numbers, we see that the faster car has more momentum than the railroad car. Now compare the faster car to an electron:

$$\mathbf{p}_{car} = m\mathbf{v} = (2{,}000 \text{ kg})(70 \text{ mi}/\cancel{h})\left(\frac{1 \cancel{h}}{3{,}600 \text{ s}}\right)\left(\frac{1{,}609 \text{ m}}{1 \cancel{mi}}\right) = 62{,}572 \text{ kg} \cdot \text{m/s}$$

$$\mathbf{p}_{electron} = m\mathbf{v} = (9.11 \times 10^{-31} \text{ kg})(2 \times 10^6 \text{ m/s}) = 1.8 \times 10^{-23} \text{ kg} \cdot \text{m/s}$$

Exercise 6.2

1. C Force is impulse divided by time. The second case exerts twice the force, as shown:

$$\mathbf{F}_1 = \frac{m(\mathbf{v}_f - \mathbf{v}_o)}{t} = \frac{m(-10-10)}{t} = \frac{-20 \, m}{t}$$

$$\mathbf{F}_2 = \frac{m(\mathbf{v}_f - \mathbf{v}_o)}{t} = \frac{m(-20-20)}{t} = \frac{-40 \, m}{t}$$

2. A This is the definition of impulse.

3. C Force is change in momentum divided by time. Area is force times time, or impulse.

4. D First, a ball that bounces somewhat elastically exerts twice the force, because the velocity goes from positive to negative, which doubles the force of an object that sticks and comes to a stop. Second, hitting near the top provides more torque on the object above its center of mass. An example of this would be car bumpers: A bumper that crumbles will exert far less force on occupants that one that bounces.

5. B Force is change in momentum divided by time. Area is force times time, or impulse.

6. A Force is change in momentum divided by time. If the change in momentum is constant and time is 1/2 as much, then force is twice as much.

7. A Force is change in momentum divided by time. As time increases, force decreases.

8. C $\mathbf{F} = m\Delta \mathbf{v}/t$ $\mathbf{F}t = m\Delta \mathbf{v}$ If contact time is the same, if $\mathbf{v}$ is doubled, then $\mathbf{F}$ is doubled.

Exercise 6.3

1. C Assign motion to the right as positive and motion to the left as negative.

$$\mathbf{p}_{before} = \mathbf{p}_{after}$$
$$m_1\mathbf{v}_1 + m_2\mathbf{v}_2 = m_1\mathbf{v}_{1f} + m_2\mathbf{v}_{2f}$$
$$(2 \text{ kg})(1 \text{ m/s}) + (1 \text{ kg}\,(-3 \text{ m/s}) = (2 \text{ kg})(-2 \text{ m/s}) + (1 \text{ kg})\mathbf{v}_{2f}$$
$$\mathbf{v}_{2f} = 3 \text{ m/s}$$

2. D Remember that in a totally inelastic collision the objects stick together. Assign positive direction to the right and then cancel the mass m.

$$\mathbf{P}_{before} = \mathbf{P}_{after}$$
$$m_A \mathbf{v}_A + m_B \mathbf{v}_B = (m_A + m_B)\mathbf{v}_f$$
$$\cancel{m}(2\mathbf{v}) + \cancel{m}(-4\mathbf{v}) = (2\cancel{m})\mathbf{v}_f$$
$$\mathbf{v}_f = -\mathbf{v}$$

3. C In an elastic collision, the cart's incoming velocity of +10 m/s results in a rebound velocity of −10 m/s. Now apply conservation of momentum to the collision of this cart with the cart coming in at +20 m/s. Assign positive direction to the right.

$$\bar{\mathbf{P}}_{before} = \bar{\mathbf{P}}_{after}$$
$$m_1 \mathbf{v}_1 + m_2 \mathbf{v}_2 = m_1 \mathbf{v}_{1f} + m_2 \mathbf{v}_{2f}$$
$$(\cancel{m})(-10 \text{ m/s}) + (\cancel{m})(+20 \text{ m/s}) = (\cancel{m})(\mathbf{v}_{1f}) + (\cancel{m})\mathbf{v}_{2f}$$
$$+10 \text{ m/s} = \mathbf{v}_{1f} + \mathbf{v}_{2f}$$

Velocities of 15 m/s and −5 m/s will work.

4. C Make the stationary object's initial velocity zero and assign positive direction to the right.

$$\bar{\mathbf{P}}_{before} = \bar{\mathbf{P}}_{after}$$
$$m_1 \mathbf{v}_1 + m_2 \mathbf{v}_2 = m_1 \mathbf{v}_{1f} + m_2 \mathbf{v}_{2f}$$
$$(\cancel{m})(2 \text{ m/s}) + (\cancel{m})(0) = (2\cancel{m})(\mathbf{v}_f)$$
$$\mathbf{v}_f = 1 \text{ m/s}$$

5. C The first coin transfers all its momentum to the identical stationary coin if they hit "head-on" in an elastic collision. The original coin had momentum only in one direction, so the final coin only has momentum in that same direction.
(Note: If the first coin does not hit the second coin along their centers of mass, both coins will go off at angles.)

6. D Use conservation of momentum and assign the positive direction to the right.

$$\bar{\mathbf{P}}_{before} = \bar{\mathbf{P}}_{after}$$
$$m_1 \mathbf{v}_1 + m_2 \mathbf{v}_2 = m_1 \mathbf{v}_{1f} + m_2 \mathbf{v}_{2f}$$
$$(0.5 \text{ kg})(1 \text{ m/s}) + (1 \text{ kg})(-2 \text{ m/s}) = (0.5 \text{ kg})(-2 \text{ m/s}) + (1 \text{ m/s})(\mathbf{v}_{2f})$$
$$\mathbf{v}_{2f} = -0.5 \text{ m/s}$$

7. C Use conservation of momentum and assign the positive direction to the right.

$$\bar{\mathbf{P}}_{before} = \bar{\mathbf{P}}_{after}$$
$$m_1 \mathbf{v}_1 + m_2 \mathbf{v}_2 = (m_1 + m_2)\mathbf{v}_f$$
$$(\cancel{m})(4\mathbf{v}) + (\cancel{m})(-2\mathbf{v}) = (2\cancel{m})(\mathbf{v}_f)$$
$$\mathbf{v}_f = \mathbf{v}$$

8. A Total kinetic energy of a system after an inelastic collision is not the same as total kinetic energy before the collision, although momentum is still conserved.

9. C Both linear momentum and kinetic energy are conserved in an elastic collision.

10. A During inelastic collisions, kinetic energy is converted to other forms of energy, such as sound.

Chapter 7

Exercise 7.1

1. A Amplitude of an oscillator does not affect its frequency or period. The oscillator will continue to keep the same frequency until the moment it stops.

2. A $T = 2\pi\sqrt{\dfrac{m}{k}}$ Period depends directly on mass but not on length.

3. C $T = 2\pi\sqrt{\dfrac{L}{g}}$ If length is multiplied by 4, square root of 4 is 2, so period will be doubled.

4. C A simple harmonic oscillator has a restoring force directly proportional to the displacement of the oscillator. For a pendulum, this is very close for angles less than about 10 to 15°. However, the restoring force for a pendulum is actually a sine function, which is not linear with displacement.

5. B $\mathbf{F} = -kx$ $F = mg$ $mg = kx$ $(0.1\ \text{kg})(9.8\ \text{m/s/s}) = (k)(0.1\ \text{m})$ $k = 9.8\ \text{N/m}$

Exercise 7.2

1. B $x = A\sin 2\pi ft$ $x = (1.5\ \text{m})\sin 18.8\ t$ $2\pi f = 18.8$ $f = 3\ \text{Hz}$
2. B $x = A\sin 2\pi ft$ $x = (1.5\ \text{m})\sin 18.8\ t$ $A = 1.5\ \text{m}$
3. A Since the equation is a sine function, $A = 0$ when $t = 0$ (it starts at the origin).
4. D It just changes sine in the equation to cosine, but all other values stay the same.
5. C Change in amplitude does not change the frequency of a simple harmonic oscillator.

Exercise 7.3

1. B At $t = 0.5$ s, the amplitude is maximum. At all other choices, amplitude is zero.
2. C Kinetic energy is maximum as the oscillator passes through the equilibrium position (where amplitude is zero). At $t = 1.75$ s, $A = 0$, so potential energy is a minimum and kinetic energy is maximum.
3. D Use the conversion of spring potential energy to kinetic energy.

   $\Delta PE = \Delta KE$

   $\frac{1}{2}kx^2 = \frac{1}{2}mv^2$

   $k(0.1\ \text{m})^2 = (0.1\ \text{kg})(10\ \text{m/s})^2$

   $k = 1{,}000\ \text{N/m}$

4. D Some energy is lost to other forms such as thermal due to friction, but the frequency stays the same.
5. B $\mathbf{F} = kx$, and the amplitude is 10 cm. The force is maximum each time the oscillator has the maximum displacement, which is 10 cm on either side of the equilibrium position.

Exercise 7.4

1. D Just as light waves slow down when they travel from air into water, when waves move from a faster medium to a slower medium, they slow down. Since $v = f\lambda$, the wavelength becomes shorter also. Frequency does not change, since it identifies the type of wave.

2. B Longitudinal waves oscillate in the same direction as the direction of propagation of the waves.

3. A $v = f\lambda = (10 \text{ Hz})(0.2 \text{ m}) = 2 \text{ m/s}$

4. B All sound waves are longitudinal.

5. A $v = f\lambda$ $\lambda = v/f = 340/5{,}000 = 0.068 \text{ m or } 6.8 \text{ cm}$

Exercise 7.5

1.   All the waves start at the speakers and travel outward in every direction at the same speed. Point A is a "loud point," or a point where there is constructive interference. Waves from the source on the left are exactly three wavelengths from A, and waves from the source on the right are exactly four wavelengths from point A. Those sets of waves will reach A in the same phase and will reinforce each other. At point B, waves from the left source are five wavelengths away, but waves from the right source are 1.5 wavelengths. Those waves will reach B in opposite phases (i.e., crest to trough) and destructively interfere—effectively canceling the sound.

2.   At a point halfway between the two speakers, the distance each wave travels from the speaker is the same, so as the waves are produced by the speakers, they will constructively interfere and produce a loud sound—as long as the speakers are set to produce sounds in the same phase at the same time.

Exercise 7.6

1. A Since there are 10 beats, the frequencies are 10 hertz apart. The second sound is $220 \text{ Hz} \pm 10 \text{ Hz}$. The second tube is shorter, so it will produce a sound of higher frequency, so the frequency of the second tube is $220 \text{ Hz} + 10 \text{ Hz}$.

2. C There are 4 beats per second, so the frequencies are 4 Hz apart. The shorter tube has a higher frequency, so $f = (260 + 4) \text{ Hz}$.

3. C Speed of sound will increase, but length of tube won't appreciably, so wavelength stays the same. $v = f\lambda$, so as speed increases with temperature, frequency or "pitch" will also increase. (Instruments will sound flat if they're not warmed up first.)

4. B Intensity is proportional to the inverse of distance from a sound. This is an "inverse square law." When distance is doubled, intensity drops to one-fourth.

5. C For a tube open at both ends, the wavelength is approximately two times the length of the tube. So the wavelength here is 3.0 m. $v = f\lambda$, so $f = v/\lambda = (340 \text{ m/s})/(3 \text{ m})$ $f = 113 \text{ Hz}$.

Chapter 8

Exercise 8.1

1. D Moment of inertia is a scalar quantity that is a property of a system.

2. C Find total kinetic energy by adding translational and rotational:

$$\text{KE} = \tfrac{1}{2}mv^2 + \tfrac{1}{2}I\omega^2 = \tfrac{1}{2}(1 \text{ kg})(2 \text{ m/s})^2 + \tfrac{1}{2}\left[\tfrac{2}{5}(1 \text{ kg})(0.2 \text{ m})^2\right]\frac{(2 \text{ m/s})^2}{(0.2 \text{ m})^2} = 2.8 \text{ J}$$

3. D Use $v = \omega R$ and convert units:

$$\omega = (45 \text{ rev}/\text{min})(2\pi \text{ rad}/\text{rev})(1 \text{ min}/60 \text{ sec}) = 4.7 \text{ rad/sec}$$
$$v = \omega R = (4.7 \text{ rad/sec})(0.08 \text{ m}) = 0.38 \text{ m/s}$$

4. D ¼ of the circumference = ¼ $(2\pi R)$ = ¼ $(2\pi)(20 \text{ m})$ = 31.4 m   angle = ¼ (2π) = $\pi/2$ rad

5. A π radians = ½ $(2\pi R)$ = πR = $\pi(20 \text{ m})$ = 62.8 m

6. B $\omega = \theta/t = (5)(2\pi \text{ rad})/1.25 \text{ sec}$ = 25 rad/sec

7. A Every point on the wheel has the same angular velocity, but points farther from the axis of rotation have a greater linear velocity. Outer points have a greater distance to travel in the same time.

8. C $v = \omega R = (\pi/15 \text{ rad/sec})(3 \text{ m})$ = $\pi/5$ rad/sec

9. A Their angular displacements are the same, since they each ran half the circle, or 180°. Time for each is not the same, so their linear velocities are not the same—they are greater when time is less.

Exercise 8.2

1. C Torque = RF, with radius perpendicular to force. Make counterclockwise positive.

A: Torque = $(20 \text{ N})(10 \text{ m})$ = +200 N-m

B: Torque = $(50 \text{ N})(2 \text{ m})$ = +100 N-m

C: Torque = $(30 \text{ N})(10 \text{ m})$ = −300 N-m

D: Torque = $(10 \text{ N})(10 \text{ m})$ = −100 N-m

2. A Add all above torques to get 100 N-m clockwise

3. B Rotational inertia is a scalar quantity.

4. B The thin rod indicates that the rod has negligible mass. The ratio of the masses on the ends is 300/200, or 1.5:1, so for torques to be equal, the ratio of distances needs to be the same but reversed. The stick is 100 cm long, so the distances 60/40 will work.

Torque clockwise = Torque counterclockwise
(300)(40) = (200)(60)

5. D Torque due to force **F** is $(100 \text{ N})(0.20 \text{ m})$ = 20 N-m, so the torque due to friction force needs to be 20 N-m counterclockwise.

6. B Torque = $FR = mgR = (50 \text{ kg})(9.8 \text{ m/s/s})(0.18 \text{ m})$ = 88.2 N-m

7. B Torque = $FR = mgR = (5 \text{ kg})(9.8 \text{ m/s/s})(0.3 \text{ m})$ = 14.7 N-m

8. A Assume the mass of the rod is exerted at its center, so there are two torques clockwise. Clockwise torque is assigned a negative.

Torque counterclockwise = $(0.2 \text{ kg})(9.8 \text{ m/s}^2)(0.2 \text{ m}) + (0.5 \text{ kg})(9.8 \text{ m/s}^2)(0.7 \text{ m})$ = 3.82 N·m

Torque clockwise = $-(1 \text{ kg})(9.8 \text{ m/s}^2)(0.3 \text{ m})$ = −2.94 N·m

Net torque = 0.88 N·m

9. D Net torque and net force on a system must both be equal to zero for a system to be in equilibrium. The system can be stationary or moving at a constant velocity.

10. D Masses 1 and 2 cause clockwise torque and mass 3 causes counterclockwise torque, so the torques due to 1 and 2 must balance the torque due to 3. In answer choice D, torque 1 is $(1 \text{ kg})(g)(2 \text{ m})$ and torque 2 is $(2 \text{ kg})(g)(2 \text{ m})$ and torque 3 is $(3 \text{ kg})(g)(2 \text{ m})$. The sum of first two balances the third.

Exercise 8.3

1. **D** Shifting the position of the axis of rotation changes the moment of inertia.

2. **A** Rotational inertia decreases due to decrease in radius, so angular velocity increases due to conservation of angular momentum.

$$I(\text{original}) \times \omega(\text{original}) = I(\text{final}) \times \omega(\text{final})$$

3. **C** Angular velocity increases because the decreasing radius decreases the rotational inertia. Angular momentum is conserved, since $L = I\omega$.

4. **D** Angular momentum is the same before and after if there is no net external torque applied to the system.

5. **C** When the radius is reduced by ½, rotational inertia is reduced by ¼, since the radius is squared in the formula: $I = kmr^2$. If the final rotational inertia is ¼ as much, the final angular velocity must be four times as much:

$$I_1\omega_1 = I_2\omega_2$$
$$(0.4 \ mr^2)(16 \text{ rad/s}) = (0.4 \ m(\tfrac{1}{2}r)^2)(64 \text{ rad/s})$$

6. **B** Shifting the axis of rotation so that it is farther from the center of mass of the object causes the moment of inertia to increase. In the case of the stick, its moment of inertia is $\frac{1}{12}mr^2$, with the axis at the center, and $\frac{1}{3}mr^2$ with the axis at one end.

Chapter 9

Exercise 9.1

1. **A** The object's density is 0.357 times the density of water. Water's density is 1,000 kg/m³, so the object's density is 357 kg/m³.

2. **B** The rock's mass is 12 g and volume is 6 cubic centimeters since that is the volume of water displaced by the object when it sinks. Density equals mass/volume = 12 g/6 ml.

3. **B** When the block sinks 60% in water, it means it is 60% as dense as water. The block's density is 600 kg/m³. The alcohol has a density of 900 kg/m³, so the block is 600/900 as dense as the alcohol and will sink in that ratio below the surface—or 2/3 below.

4. **D** Multiply by 1,000 to convert g/ml to kg/m³.

5. **A** Density equals mass/volume = 10 g/(2 cm)³ = 10/8 = 1.25 g/ml or 1.25 g/cm³

Exercise 9.2

1. **C** A depth of 6 m of water is about 0.6 atm, since as a general rule 10 m depth of is equal to 1 atm ($P = \rho gh$). Therefore, the total pressure is 1 atm of air pressure plus 0.6 atm of water pressure.

2. **C** Pressure only depends on the depth of fluid and does not depend on amount of fluid or surface area. $P = \rho gh$

3. **A** From Pascal's Principle, pressure at the bottom is the same, since pressure is the same in all directions at a given depth in a fluid.

4. **D** Area does not affect fluid pressure at a depth. Depth of 15 m in water is equal to about 1.5 atm, so the total pressure is 1.5 atm of water plus 1 atm of air pressure. That total of 2.5 atm is multiplied by 1.03×10^5 Pa per atm.

5. **C** $P(\text{total}) = 1 \text{ atm (air)} + \rho gh \text{ (seawater)} = 101,000 \text{ Pa} + (1.025)(9.8)(4,000)\text{Pa}$
$P = F/A$, so $F = PA = (40,281,000 \text{ Pa})(1 \text{ m}^2) = 40,281,000 \text{ N}$.

6. D Fluid pressure at a depth does not depend on surface area or total volume—only on depth.

Exercise 9.3

1. A Buoyant force upward equals gravitational force downward.

2. B The object displaces the same volume as its own volume when it sinks, but that same volume of seawater weighs more than fresh water. So the buoyant force is greater and the object will "lose" more of its actual weight if it sinks or float higher in seawater than in fresh water.

3. C At the bottom, the forces are normal force and buoyant force upwards and gravitational force downward. An object sitting on the bottom still has a buoyant force on it.

4. D A mass of 100 g is equal to (0.1 kg)(9.8 m/s/s) = 0.98 N in air.
It displaces (0.05 g)(9.8 m/s/s), or 0.49 N of water.
So the buoyant force is 0.49 N and the apparent weight is 0.98 N − 0.49 N, or 0.49 N.

5. C It will have to displace its own weight, which is (0.1 kg)(9.8 m/s/s) = 0.98 N.
Water displaced must be 0.98 N = mg, so mass of water displaced is 0.1 kg.
A mass of 0.1 kg of water is 100 ml, which has a volume of 100 ml.

6. D The buoyant force is equal to the weight of water displaced.
40 L of water = 40,000 ml = 40,000 g = 40 kg
F(buoyant) = (40 kg)(9.8 m/s/s) = 392 N

7. B If the ball floats, the buoyant force is equal to the weight of the ball. 10 g = 0.01 kg
F(buoyant) = (0.01 kg)(9.8 m/s/s) = 0.098 N

8. B Buoyant force is equal to the density of the fluid displaced, the volume of fluid displaced, and g.

9. C There is no buoyant force at all on the Moon, since the Moon has no atmosphere to provide a surrounding fluid to generate the buoyant force. There is gravity, but no upward force other than normal force at the surface.

10. C The forces are buoyant force upward, gravitational force downward, and tension upward.

11. B The forces are two tension forces upward (2T) and buoyant force upward and gravitational force downward.

12. D The weight of the ball is equal to the weight of fluid displaced when it floats.

Exercise 9.4

1. A Continuity principle: When the spout becomes narrow, the water increases speed. The faster moving water strikes a surface with greater change in momentum and thus greater force.

2. C By cutting the radius in the area calculation to ¼r and squaring, we get an area that is 1/16 as much, so velocity is 16 times: $A_1v_1 = A_2v_2$

3. D The continuity equation $A_1v_1 = A_2v_2$ assumes density remains the same, so it has been canceled from both sides of the equation. If density is included, the units are kilograms per second, which is mass rate of flow.

4. D Speed of flow increases, so the amount of water flowing per second remains the same.
Q = volume/time = Av

5. B $Q = \dfrac{v}{t} = Av$ Q = volume = Avt = (½ πr^2)(3)(3600) m³ = 678 m³

Exercise 9.5

1. A Bernoulli's equation is a conservation of energy statement.

2. A When a fluid moves faster, its fluid pressure decreases, according to Bernoulli's equation and conservation of energy principles.

3. B According to the continuity principle, $A_1 v_1 = A_2 v_2$, fluid moves faster in the thinner tube. Bernoulli's principle then states that when the fluid moves faster, its pressure is lower.

4. C The change in kinetic energy of the water from the ground level is equal to the gain in potential as it reaches its maximum height.

5. A Potential and kinetic energy take this form in fluids: $\rho g h = \frac{1}{2} \rho v^2$

Chapter 10

Exercise 10.1

1. C Heating or heat is the transfer of energy from a higher temperature to lower temperature.

2. A Thermal energy is energy within a system, whereas heat is the transfer of energy.

3. C Temperature is a measure of the kinetic energy of molecules, so $T \propto \text{KE}$.

4. D Temperature is proportional to kinetic energy and kinetic energy is proportional to square of velocity, so temperature is proportional to square of velocity. If temperature is doubled, then velocity is the square root of two (which would be doubled when velocity is squared). $2T \propto (\sqrt{2}\,v)^2$

5. B If temperature doesn't change and volume is decreased, pressure increases because molecules collide with the sides of the container more frequently. If there is no change in temperature, there is no change in average kinetic energy of the molecules.

6. A If temperature increases and volume stays constant, the speed of molecules increases and molecules collide with walls of the container at higher speeds. The collisions have greater change in momentum and exert more force—and more pressure.

7. B Add the change in length to the original length to get the new length. It is a very small change, but even a small change can cause buckling in the bridge if spaces aren't built into the original bridge.

$$L = L_o + \Delta L = L_o + \alpha L_o \Delta T = 30 \text{ m} + (30 \text{ m})(7.2 \times 10^{-6}/\text{C}°)(40°\text{C}) = 30.009 \text{ m}$$

Exercise 10.2

1. A $PV = nRT$. Pressure increases as temperature increases if volume is constant.

2. B $PV = nRT$. Volume is constant and temperature is doubled, so pressure increases and internal energy increases. However, if there is no change in volume, there is no work done, since $W = -P\Delta V$.

3. D A, B, and C are all true statements.

4. A Double the number of molecules means double the frequency of collisions.

5. C Ideal gases are assumed to have elastic collisions.

6. C At a depth of 21 m, the total fluid pressure is 2 atm of water plus 1 atm of air. As the bubble goes to the surface where $P = 1$ atm, the pressure reduces to 1/3, so the bubble's volume is three times as much. At constant temperature, $P_1 V_1 = P_2 V_2$.

7. A $\frac{P_1}{T_1} = \frac{P_2}{T_2}$ Pressure increases as absolute temperature increases.

Exercise 10.3

1. C A shorter rod decreases L and thus increases the rate of heat flow.

$$Q = \frac{kA\Delta T}{L}$$

2. C Energy is conserved, so heat flow out equals heat flow in to two connected systems.

3. A $W = -P\Delta V$ In choice A, ΔV can be decreased, so the negative sign means work is positive.

    In choice B, isovolumetric means $\Delta V = 0$, so no work.

    In choice C, internal energy won't change if Q out equals work.

    In choice D, isothermal means $\Delta T = 0$, so internal energy won't change, since internal energy depends on temperature.

4. A Isothermal means $\Delta T = 0$, so internal energy won't change and KE won't change.

5. B Adiabatic means no heat transfer in or out of the system.

Chapter 11

Exercise 11.1

1. A The charge on an electron is 1.6×10^{-19}, so 1×10^{11} electrons would have a total charge of 1.6×10^{-8} coulombs.

2. C Total charge is $-2Q$, and it will distribute evenly between the two spheres, with $-Q$ on each sphere.

3. A The negative rod induces a charge separation or polarization on the second rod, with positive charges closer to the negative rod. If they don't touch, there is no charge transfer, and the second rod will return to an even distribution of charge and remain neutral.

4. B The second rod was polarized when the first (negative) rod came near it, with the near end more positive and far end more negative. When the grounding wire was attached to the second rod, negative charges then left the rod, leaving it with a net positive charge. When the grounding wire was cut, the charge remained when the first rod was removed.

5. B All the charges repel each other and are able to move freely on the sphere, so they distribute on the surface evenly.

Exercise 11.2

1. A The field is to the right from both charges—outward from the positive and inward toward the negative—so their magnitudes add to each other.

$$E = \frac{kQ}{r^2} = (9 \times 10^9)\left(\frac{2 \times 10^{-6}}{4} + \frac{4 \times 10^{-6}}{9}\right) = 8{,}500 \text{ N/C}$$

2. B The charge on the right is of larger magnitude, so the field from it can "reach" to the left of the 2-μC charge, where the fields from the two charges would be in opposite directions and add to zero.

3. C The field is downward from positive to negative plates; electric potential increases upward from negative to positive plate; a proton would be attracted to the negative plate and experience a force downward.

4. B Both charges have fields outward. At the origin, the fields are in opposite directions and equal to each other due to the symmetry of the situation, so they would cancel to zero. At the point P, the fields from both will cancel in the x-direction and add in the y-direction—upward.

5. C $E = \dfrac{kq}{r^2} = \dfrac{(9 \times 10^9)(2 \times 10^{-9})}{2^2} = 45,000$ N/C

6. D The electric field between charged plates is uniform when the plates are parallel and magnitudes of the charges are equal.

7. B Both charges cause a field to the right—outward from the positive on the left and inward toward the negative on the right.

Exercise 11.3

1. D $F = \dfrac{kQq}{r^2}$. If both charges are doubled, that would be four times the value on the right, so the force will be four times as much.

2. A $F = \dfrac{kQq}{r^2}$. If the radius is four times as much, it is then squared, so the force would be 1/16 of its value. $F/16 = 0.00125$ N.

3. D Zero. Due to the symmetry of the situation, the four charges are all the same distance from the center, so they exert the same four vector forces that are directed so they cancel.

4. D If the gravitational force is negligible, the charge moving between the plates experiences a single electrical force in one direction (similar to a ball moving through the air), so the resulting path is parabolic. Once the charge leaves the edge of the plates, it will take a linear path, since there will be no electrical force on it. (Remember, we are neglecting the gravitational force, since it is so small in comparison.)

5. C The positive charge attracts the negative to the left.

$$F = \dfrac{kQq}{r^2} = \dfrac{(9 \times 10^9 \text{ N} \cdot \text{m}^2/\text{C}^2)(3 \times 10^{-6} \text{ C})(4 \times 10^{-6} \text{ C})}{(5 \text{ m})^2} = 0.0043 \text{ N}$$

6. A The net force is to the right. The positive charge at $x = 1$ repels the proton to the right, and the negative charge at $x = 6$ attracts the proton to the right.

7. A The numerator is twice as large, but the denominator is twice as large but squared. So the new value is ½ as much.

8. D $F = \dfrac{kQq}{r^2} = \dfrac{(9 \times 10^9 \text{ N} \cdot \text{m}^2/\text{C}^2)(5 \times 10^{-6} \text{ C})^2}{(4 \text{ m})^2} = 0.014 \text{ N}$

9. D $F = \dfrac{kQq}{r^2} = \dfrac{(9 \times 10^9 \text{N} \cdot \text{m}^2/\text{C}^2)(1.6 \times 10^{-19} \text{ C})^2}{(1 \times 10^{-9} \text{ m})^2} = 2.3 \times 10^{-10} = 0.23 \times 10^{-9} \text{ N} = 0.23 \text{ nN}$

Exercise 11.4

1. D Electric potential is a scalar, so the two positive potentials will add to the two equal negative potentials (equal charge magnitudes at equal distances) to produce zero.

2. A All the charges are positive, so all the potentials will be positive and add to a positive potential.

3. B All the charges are negative, so all the potentials will be negative and add to a negative potential.

4. A $V = \dfrac{kq}{r}$. The one that is twice as far will have half the potential.

5. C $U = \dfrac{kQq}{r} = \dfrac{(9 \times 10^9)(5 \times 10^{-6})(2 \times 10^{-6})}{0.004} = 22.5 \text{ J}$

6. B $U = \dfrac{kQq}{r} = \dfrac{(9 \times 10^9)(10 \times 10^{-6})(1.6 \times 10^{-19})}{6} = 2.4 \times 10^{-15} \text{ J}$

Chapter 12

Exercise 12.1

1. A $I = \dfrac{\Delta Q}{\Delta t} = \dfrac{3.6 \times 10^{-6}}{10 \times 10^{-3}} = 0.36 \text{ mA}$

2. C $1 \text{ A} = 1 \text{ C/s}$

3. B Conventional current is counterclockwise in the circuit, from the high-potential (longer bar) side of the battery, so the current moves to the right in that resistor. More current will flow into the branch with the least resistance.

4. C The current from the battery splits when it reaches the parallel junction, so only part of the 10-A current goes through each branch. The total current in the two branches has to be 10 A.

5. B Conventional current is the direction positive charge would flow, which is from high potential (+) to low potential (−). Electron flow is from low potential (−) to high potential (+).

Exercise 12.2

1. C R(series) = 30 ohms R(parallel) = product/sum = (300)(300)/600 = 150 Ω

2. B Only the two in series: 100 Ω + 200 Ω

3. C The 200 and 300 are in parallel. R(total) = product/sum = (200)(300)/500 = 120 Ω

4. D R = product/sum = (500)(300)/800 = 187.5 Ω. Resistance is greater, so current is less.

5. C $R = \dfrac{\rho L}{A}$; $R(\text{new}) = \dfrac{\rho(\frac{1}{2}L)}{\pi(2r)^2} = \frac{1}{8}R$.

Exercise 12.3

1. D $U = \frac{1}{2}CV^2$. If the voltage is cut from 20 V to 10 V, since voltage is squared, the energy stored will be cut from 100 μJ to 25 μJ.

2. B $U = \frac{1}{2}CV^2 = \frac{1}{2}(0.002)(100) = 0.1 \text{ J}$

3. A Positive work is done in moving the plates apart because the plates are attracted to each other due to opposite charges, so the force to move them is outward and the displacement is also outward.

4. C $Q = CV$. If voltage is doubled, the capacitance stayd the same, so charge is doubled.

5. A Capacitors in series add like resistors in parallel, so we can use the product/sum rule. C(total) = (2,000 μF)(4,000 μF)/6,000 μF = 1,300 μF.

6. A For capacitors in parallel, add them as you would resistors in series.

Exercise 12.4

1. D R(total) = 150 ohms $I = V/R = 6\ V/150\ \Omega = 0.04\ A$.

2. D The current splits evenly in the two branches because total resistance in each branch is the same (150 Ω in each branch). The current through the two resistors in series is the same.

3. C $I = 0.04\ A = 40\ mA$. The current in each resistor in the parallel branches is 20 mA.

4. D The two resistors are in series, so add their resistances to find the total, which is 50 Ω. Use Ohm's law to find current: $I = V/R = 10\ v/50\ \Omega = 0.2\ A$. This current is the same in both resistors.

5. C On a plot of potential difference versus current, the slope will be resistance, since $V = IR$.

Exercise 12.5

1. C An ammeter reads current, which is the same everywhere in a series circuit.

2. B The total resistance is the total for the resistors in the parallel branches plus the resistor connected in series. R(parallel) = product/sum = $(400)(600)/1,000\ \Omega = 240\ \Omega$
   R(total) = 240 + 100 = 340 Ω

   $I = \Delta V/R = 20\ V/340\ \Omega = 0.059\ A$ leaving the battery, with all going through the 100 Ω.

3. D The current splits proportionally, with 60% in the 400 Ω and 40% in the 600 Ω. More current takes that path with least resistance.

4. A The current is the same leaving the battery and returning to the battery.

5. C The current leaving the battery is equal to the sum of currents splitting into the parallel branches and then the total returns to the battery.

Exercise 12.6

1. B Using Kirchhoff's loop rule, the potential difference (voltage) across the battery is equal to the sum of voltages across the other two resistors in a series circuit; i.e., the voltage supplied by the battery is equal to the sum of voltages used by the resistors. So the total change in energy (using V = energy/charge) for a circuit loop is zero.

2. B ΔV(battery) = ΔV(resistor 1) + ΔV(resistor 2)

3. B R(total) = 340 Ω and $I = 0.3$ amp. For the 100-ohm resistor: $\Delta V = IR = (0.3)(100) = 30\ V$

4. A Kirchhoff's loop rule: battery voltage – resistor 1 voltage = resistor 2 voltage

5. D Each parallel branch is in a loop with the battery and the 100-ohm resistor. So the battery voltage minus voltage across the 100-ohm resistor is the voltage in each of the parallel branches. This will always work so that we can state: "voltages across parallel branches are equal."

Exercise 12.7

1. B Total resistance is the sum of the two resistors in series combined in parallel with the third resistor. $R = \dfrac{(200)(100)}{300} = 67\ \Omega$ $I = \dfrac{10\ V}{67\ \Omega} = 0.015\ A$

2. C $P = VI = (10\ V)(0.015\ A) = 1.5\ W$

3. C $P = I^2R$ Resistor A gets 2/3 of the current, which is 0.1 A.

                    Power of A = $(0.1 \text{ A})^2(100 \text{ }\Omega) = 1$ W

                    Resistors B and C get 1/3 of the current in their branch, which is 0.05 A in B and in C. Power of B = Power of C = $(0.05 \text{ A})^2(100 \text{ }\Omega) = 0.25$ W each

Notice that the total power produced by the battery in Question 2 is the same as the total power consumed by the three bulbs. That will always be true.

4. C $P = VI = (19.5)(3.34) = 65$ W

    $E = Pt = (65 \text{ W})(1800 \text{ sec}) = 117{,}000$ J or 1.2×10^5 J

5. A $P = E/t$, so $E = Pt = (40 \text{ W})(60 \text{ h}) = 2{,}400$ W-h or 2.4 kWh

Chapter 13

Exercise 13.1

1. A Use the right-hand rule, with thumb pointing in direction of current and fingers curled around the wire.

2. B $B = \dfrac{\mu_o I}{2\pi R} = \dfrac{(4 \times 10^{-7})(3)}{2\pi(0.2)} = 3 \times 10^{-6}$ T

3. C $B = \dfrac{\mu_o I}{2\pi R}$ If I is doubled, B is also doubled.

4. A $B = \dfrac{\mu_o I}{2\pi R}$ If R is doubled, B is cut in half.

5. D Length is not a factor in the calculation of B using this equation.

Exercise 13.2

1. B $F = qvB = (1.6 \times 10^{-19})(2 \times 10^5)(200) = 6.4 \times 10^{-12}$

    The force is out of the page, using the right-hand rule.

2. B Only the component perpendicular will generate a force, so the direction will be the same.

3. D The velocity must be perpendicular to the B field to generate a magnetic force.

4. C Electrical force is qE and magnetic force is qvB. Charges not moving in a magnetic field produce no force. Stationary charges in an electric field do experience a force. The force in B is twice the force in A, but the force in C is the largest.

5. D There is no force since the charge is not moving.

6. A The electron is negative, so the force is opposite the direction of the electric field. $F = qE$

7. D There must be a perpendicular component of velocity to produce a force in a magnetic field.

8. A Right-hand rule: Velocity is to the right, and B is into the page, so F is upward on the page.

9. B The force keeps changing direction as the particle changes direction, so it becomes a circular path with the magnetic force providing the centripetal force.

10. B The magnetic force is upward on the page, so a force is needed down on the page. For a positive charge, this would be an electric field down on the page: $F = qE$

Exercise 13.3

1. B The plane of the loop must be perpendicular to the field so the magnetic field in the loop is the same direction as the area vector to create flux.

2. B Right-hand rule: Thumb points along the current (to the right), so fingers curl around the wire and out of the page above the wire on the page.

3. C $F = ILB$ Find B first and substitute.

$$\mathbf{F} = IL\left(\frac{\mu_o I}{2\pi r}\right) = \frac{(0.5 \text{ A})(1 \text{ m})(4\pi \times 10^{-7})(0.5)}{2\pi(0.002)} = 2.5 \times 10^{-5} \text{ N} \quad \text{up on the page}$$

4. D Use the right-hand rule to find out of the page. Both wires have current to the right, so curl the fingers and they come out of the page above the wire.

5. C When the current is switched, each wire creates a field in the opposite direction—one into the page and one out. However, the closer wire creates the stronger field.

Chapter 14

Exercise 14.1

1. C Frequency defines the type, or "color" of the radiation because speed and wavelength can change in a medium, but frequency does not.

2. B The object reflects blue light, but the yellow filter allows only red and green light to pass, so the object appears black through the filter.

3. B This is the correct ranking. The lowest frequency has the longest wavelength.

4. C $v = f\lambda$ $f = v/\lambda = (3 \times 10^8 \text{ m/s})/(400 \times 10^{-9} \text{ m}) = 7.5 \times 10^{14}$ Hz
   The other one is 4.3×10^{14} Hz.

5. B 1,000 nm is longer than visible wavelength range.

6. A $n = 1.3 = c/v = \lambda(\text{air})/\lambda(\text{medium})$ $\lambda = 624$ nm/1.3 = 480 nm

7. C The shirt reflects all colors. The yellow filter allow red and green through, and the magenta filter allows blue and red. Red will pass through both.

8. C $f = v/\lambda = (3 \times 10^8)/(530 \times 10^{-9}) = 5.66 \times 10^{14}$ Hz

Exercise 14.2

1. C Use the thin-lens equation with $R = 20$ cm, $f = -10$ cm (virtual focus), and object distance 5 cm.

   $$\frac{1}{f} = \frac{1}{d_o} + \frac{1}{d_i}$$ Image distance is −3.3 cm, so it is virtual and upright.

   The ratio of image distance to object distance is 3.3/5, so the image is smaller than the object.

2. C Use the thin-lens equation with focal length 6 cm and object distance 8 cm.
   The object is outside the focal length, so it should create a real and inverted image.

   $$\frac{1}{f} = \frac{1}{d_o} + \frac{1}{d_i}$$ Image distance is 24 cm.

3. C $\dfrac{1}{f} = \dfrac{1}{d_o} + \dfrac{1}{d_i}$

$\dfrac{1}{d_i} = \dfrac{1}{f} - \dfrac{1}{d_o} = \dfrac{1}{6} - \dfrac{1}{8} = \dfrac{4}{24} - \dfrac{3}{24} = \dfrac{1}{24}$

$d_i = 24$ cm

4. B Magnification is image distance divided by object distance, which is 24/8.

5. A A plane mirror produces an image that is the same distance behind the mirror as the object is in front. The image is upright and virtual and the same size as the object. It is flipped left to right.

6. C For plane mirrors, the image is the same distance behind the mirror as the object is in front of the mirror. The image is upright, virtual, and inverted left to right.

7. A Use the thin-lens equation with focal length 6 cm and object distance 3 cm. Image distance is −6 cm, so the image is virtual and upright. The magnification is −6/3 or −2, so the image is larger (and negative sign says it's virtual).

8. A Convex mirrors do not form real images.

9. B Concave mirrors form real images if the object is placed outside of the focal length.

10. A The virtual rays are a projection by the brain of the reflected rays that are directed toward the eyes.

Exercise 14.3

1. B Ray B is shown to refract towards the thin part of the lens, which is incorrect.

2. B Project rays back to the same side as the object to see that the image will be smaller than the object. The rays would have to converge on the far side of the lens to form a real image.

3. C Virtual images are upright. This lens won't form a real image.

4. B Ray C hits the lens as it heads toward a focal point and then refracts parallel to the principal axis.

5. B $n = c/v$. If $n = 1.5$, the speed of light in air is 1.5 times the speed in the medium—or the speed in the medium is 2/3 the speed in air.

6. C Higher frequencies refract over larger angles.

7. B Critical angle for a medium: $\sin \theta_c = 1/n = 1/1.42$; $\theta_c = 44.8°$.

8. B Convex lenses form real images if object distance is greater than focal length.

9. A Concave lenses never form real images.

10. A The graph is based on Snell's law. A graph with sine angle of incidence on the y-axis and sine angle of refraction is the x-axis would have a slope equal to the index of refraction of the medium (since the index for air is just 1).

11. D $\dfrac{1}{f} = \dfrac{1}{d_o} + \dfrac{1}{d_i} = \dfrac{1}{10} + \dfrac{1}{10} = \dfrac{2}{10}$

$f = 5$ cm

12. B Light rays coming in to the lens bend toward the thickest part of the lens. In this case, the concave lens is thicker on the outside, so light bends away from the center as it passes through the lens and does not come to a focus.

Exercise 14.4

1. B   First order lines form equidistant on either side of the principal maximum at the center.

2. B $m\lambda = d \sin\theta$, where d is the spacing between openings and θ is the angle between bright lines. If d is larger, then the angle will be smaller and the bright lines will be closer.

3. B   If the wavelength gets larger, then the angle is larger and spacing between lines is larger.

4. D   All are true except D.

Chapter 15

Exercise 15.1

1. D   The weak nuclear force is a short-range force found only in the nucleus that is believed to be responsible for beta decay.

2. C   The strong nuclear force is considered to be the strongest force in the universe.

3. D   The friction force is a manifestation of the electromagnetic force, or electromagnetic interactions among molecules of both surfaces.

4. C   The strong force is a short-range nuclear force that holds nucleons together.

5. A   The gravitational force is a long-range force considered to be the weakest of the fundamental forces and is responsible for attraction of all objects in the universe.

Exercise 15.2

1. C   An electron is fundamental because it is not believed to be made up of any other particles.

2. D   Isotopes of the same element have different numbers of neutrons in the nucleus.

3. C He_2^4 has two protons in the nucleus (which identifies it as helium) and a mass number of 4, which would be 2 protons plus 2 neutrons.

4. C   Positrons have the same mass as electrons but a positive elementary charge n.

5. D U_{92}^{238} has 92 protons in the nucleus, 238 total number of nucleons, and $238 - 92$ or 146 neutrons in the nucleus

Exercise 15.3

1. C   After one half-life, a single nucleus has either decayed or it has not. It had a 50% probability of decay.

2. C   Beta β_{-1}^0 is given off, so the number of protons has increased by 1 to make the total charge on the product side of the equation equal the charge on the reactant side.

3. D   100 g reduces to 50 g after about 140 days, so this would be the half-life.

4. A   The original charge (lower total) on the left side is 84, so the total on the product side is $2 + 82$. The upper numbers account for mass. On the left side, the total is 210, so the total on the right side is $4 + 206$.

5. C   That is two half-lives, since during that time the mass decayed from 100 g to 50 g and then from 50 g to 25 g.

Exercise 15.4

1. C Longer arrows on the diagram represent larger changes in energy, so that would also be higher frequency. Arrows up represent absorption of a photon for each transition, and arrows down represent emission of a photon for each transition.

2. C A short wavelength is high frequency: $E = hf = hc/\lambda$

3. B $n = 1$ to $n = 4$ is a gain in energy, so a photon is absorbed: $\Delta E = E_f - E_o$

4. B $\Delta E = E_f - E_o = hf = \dfrac{hc}{\lambda}$.

Exercise 15.5

1. B When helium nucleus is given off, that reduces the charge by 2 and the mass number by 4, so the new product must have a lower charge by 2 and a lower mass number by 4.

2. B When a beta is given off, with a charge of -1 and mass of zero, the new product must have a charge that is 1 higher than before, but the mass does not change.

3. C Conservation of momentum.

4. D KE = 0 at $f = 1 \times 10^{15}$ Hz

5. B The total charge on the left side must be 83 since the total charge on the right side (lower numbers) is 83. The total mass on the right (upper right) must be 215 to balance mass of 215 on the left. This is conservation of mass-energy and conservation of charge.

6. C Gamma rays don't have a charge, so they don't experience a force in a magnetic field.

7. D In balancing this equation, 3 neutrons are required on the right side to balance the the total mass on the left of 235. Each neutron has a mass number of 1 and no charge.

8. C Each positron has a charge of 1 but a mass number of zero. The 4 hydrogens on the left side have a total charge of 4, so it takes 2 positrons to balance. Positrons don't have mass to affect the equation.

9. B A neutron has zero charge and a mass of 1, which is the choice that will balance the equation.

10. A The electron has gained energy during the collision, so the photon has to lose energy. Since $E = hf$, an energy decrease by the photon would be a decrease in frequency, which means wavelength increases. The photon rebounds with a longer wavelength.

11. B Electron volts (eV) are units of energy. $E = mc^2$ shows that E/c^2 will be units of mass.

Exercise 15.6

1. D $p = hf/c = h/\lambda$

2. A $p = \dfrac{h}{\lambda}$

$$\lambda = \frac{h}{mv} = \frac{6.63 \times 10^{-34} \text{ J} \cdot \text{s}}{(9.11 \times 10^{-31} \text{ kg})(1 \times 10^6 \text{ m/s})} = 7.28 \times 10^{-10} \text{ m}$$

3. C $E = hf = \dfrac{hc}{\lambda} = \dfrac{1.99 \times 10^{-25} \text{ J} \cdot \text{m}}{700 \times 10^{-9} \text{ m}} = 2.8 \times 10^{-19}$ J

Notes

Notes

Notes

Notes